CAMBRIDGE LIBRARY COLLECTION

Books of enduring scholarly value

Life Sciences

Until the nineteenth century, the various subjects now known as the life sciences were regarded either as arcane studies which had little impact on ordinary daily life, or as a genteel hobby for the leisured classes. The increasing academic rigour and systematisation brought to the study of botany, zoology and other disciplines, and their adoption in university curricula, are reflected in the books reissued in this series.

Monograph on the Fossil Reptilia of the London Clay

Covering a wide area of the London and Hampshire basins, the London Clay has been famous for over two hundred years as one of the richest Eocene strata in the country. In this work, first published between 1849 and 1858, Fellows of the Royal Society Richard Owen (1804–92) and Thomas Bell (1792–1880) describe their findings from among the reptilian fossils found there. The book is divided into four parts, covering chelonian, crocodilian, lacertilian and ophidian fossils, and includes an extensive section of detailed illustrations. Using his characteristic 'bone to bone' method and an emphasis on taxonomy, Owen draws some significant conclusions: he shows that some of Cuvier's classifications were wrongly extended to marine turtles, and adds to the evidence for an Eocene period much warmer than the present. The work is a fascinating example of pre-Darwinian palaeontology by two scientists later much involved in the evolutionary controversy.

Cambridge University Press has long been a pioneer in the reissuing of out-of-print titles from its own backlist, producing digital reprints of books that are still sought after by scholars and students but could not be reprinted economically using traditional technology. The Cambridge Library Collection extends this activity to a wider range of books which are still of importance to researchers and professionals, either for the source material they contain, or as landmarks in the history of their academic discipline.

Drawing from the world-renowned collections in the Cambridge University Library and other partner libraries, and guided by the advice of experts in each subject area, Cambridge University Press is using state-of-the-art scanning machines in its own Printing House to capture the content of each book selected for inclusion. The files are processed to give a consistently clear, crisp image, and the books finished to the high quality standard for which the Press is recognised around the world. The latest print-on-demand technology ensures that the books will remain available indefinitely, and that orders for single or multiple copies can quickly be supplied.

The Cambridge Library Collection brings back to life books of enduring scholarly value (including out-of-copyright works originally issued by other publishers) across a wide range of disciplines in the humanities and social sciences and in science and technology.

Monograph on the Fossil Reptilia of the London Clay

And of the Bracklesham and Other Tertiary Beds

Richard Owen
Thomas Bell

CAMBRIDGE
UNIVERSITY PRESS

CAMBRIDGE
UNIVERSITY PRESS

University Printing House, Cambridge, CB2 8BS, United Kingdom

Cambridge University Press is part of the University of Cambridge.
It furthers the University's mission by disseminating knowledge in the pursuit of
education, learning and research at the highest international levels of excellence.

www.cambridge.org
Information on this title: www.cambridge.org/9781108038249

© in this compilation Cambridge University Press 2015

This edition first published 1849-58
This digitally printed version 2015

ISBN 978-1-108-03824-9 Paperback

MONOGRAPH

ON THE

FOSSIL REPTILIA

OF

THE LONDON CLAY,

AND OF THE

BRACKLESHAM AND OTHER TERTIARY BEDS.

BY

PROFESSOR OWEN, D.C.L., F.R.S., F.L.S., F.G.S., &c.

AND

PROFESSOR BELL, F.R.S., F.L.S., F.G.S., &c.

LONDON:

PRINTED FOR THE PALÆONTOGRAPHICAL SOCIETY.

1849—58.

The Monograph on 'THE REPTILIA OF THE LONDON CLAY, AND OF THE BRACKLESHAM AND OTHER TERTIARY BEDS,' was published as follows:

Part I, pages 1—76, plates I—XXVIII, VIIIA, XA, XIIIA, XVIA, XVIIIA, XIX*, XIXB, XIXC, AND XIXD, July, 1849.

Part I, Supplement I, pages 77—79, plates XXVIIIA, XXVIIIB, April, 1858.

Part II, Supplement I, pages 1—4, plate XXIX, August, 1850.

Part II, pages 5—50, plates I—XII, and IIA, August, 1850.

Part III, pages 51—68, plates XIII—XVI, August, 1850.

CONTENTS

OF THE

MONOGRAPH ON THE REPTILIA OF THE LONDON CLAY, &c.

ORDER—CHELONIA. Family—MARINA.

Genus—CHELONE, observations on . .	Part I.	Page 1
Species—Chelone breviceps . . .	,,	10
,, ,, longiceps . . .	,,	16
,, ,, latiscutata . . .	,,	20
,, ,, convexa . . .	,,	21
,, ,, subcristata . . .	,,	24
,, ,, planimentum . .	,,	25
,, ,, crassicostata . .	,,	27
,, *Testudo plana* . . .	,,	27
Chelone declivis . . .	,,	30
,, ,, trigoniceps . . .	,,	31
,, ,, cuneiceps . . .	,,	33
,, ,, subcarinata . . .	,,	37
Supplemental observations on the species from Harwich	,,	40
Summary on the genus Chelone . . .	,,	44

Family—FLUVIALIA.

Genus—TRIONYX, observations on . .	,,	45
Species—Trionyx Henrici . . .	,,	46
,, ,, Barbaræ . . .	,,	50
,, ,, incrassatus . . .	,,	51
,, ,, marginatus . . .	,,	55
,, ,, rivosus . . .	,,	56
,, ,, planus . . .	,,	58
,, ,, circumsulcatus . .	,,	59
,, ,, pustulatus . . .	,,	60
,, ,, sp. ind. . . .	,,	61

CONTENTS.

Family—PALUDINOSA.

Genus—PLATEMYS, observations on		.	.	Part I.	Page 61
„	„	Bullockii .	.	„	62
„	„	Bowerbankii	.	„	66
„	EMYS	.	.	„	67
„	„	testudiniformis	.	„	67
„	„	lævis .	.	„	70
„	„	Comptoni	.	„	71
„	„	bicarinata	.	„	73
„	„	Delabechii	.	„	74
„	„	crassus	.	„	76

Family—PALUDINOSA.

Genus—EMYS CONYBEARII	.	.	Part I. Sup. 1	77
„	PLATEMYS BOWERBANKI (?)	.	„ II. „ 1	1

ORDER—CROCODILIA.

Genus—CROCODILUS, observations on	.	Part II.	5
„	„	toliapicus	„	29
„	„	champsoides	„	31
„	„	remarks on vertebræ of two preceding species	„	33
„	„	Hastingsiæ	„	37
„	„	Hantoniensis	„	42
„	„	vertebræ referable to C. Hastingsiæ	„	44
„	GAVIALIS, remarks on	.	„	46
„	„	Dixoni	„	46

ORDER—LACERTILIA, remarks on . . „ 50
„ „ Pleurodont Lizard (?) . „ 50

ORDER—OPHIDIA, remarks on . . Part III. 51

Genus—PALÆOPHIS TYPHÆUS	.	.	„	56
„	„	porcatus	„	61
„	„	toliapicus	„	63
„	„	(?) longus	„	66
„	PALERYX, remarks on	.	„	67
„	„	rhombifer	„	67
„	„	depressus	„	67

INDEX

TO THE

SPECIES DESCRIBED IN THE MONOGRAPH

ON THE

REPTILIA OF THE LONDON CLAY, &c.

Alligator Hantoniensis, *Wood*, Mid. Eo.   Part II, p. 42, tab. viii, fig. 2.

Chelone antiqua, Kœnig (?).   Part I, p. 10.

Chelone breviceps, *Owen*, Lond. Cl.   Part I, p. 10, tabulæ i, ii.

   „   convexa, *Owen*, Lond. Cl.   Part I, p. 21, tab. vii.

   „   crassicostata, *Owen*, Lond. Cl.   Part I, pp. 27 and 42, tabs. xi, xii, xiii, xiii A, and xiii B.

   „   cuneiceps, *Owen*, Lond. Cl.   Part I, p. 33, tab. xv.

   „   declivis, *Owen*, Low. Eo.   Part I, p. 30, tab. xiv.

   „   *Harvicensis, Woodward* (?).   Part I, p. 25.

   „   latiscutata, *Owen*, Lond. Cl.   Part I, p. 20, tab. vi.

   „   longiceps, *Owen*, Lond. Cl.   Part I, p. 16, tabs. iii, iv, and v.

   „   planimentum, *Owen*, Lond. Cl.   Part I, p. 25 and 40, tabs. ix, x, and x A.

   „   subcarinata, *Bell*, Lond. Cl.   Part I, p. 37, tab. viii A.

   „   subcristata, *Owen*, Lond. Cl.   Part I, p. 24, tab. viii.

   „   trigoniceps, *Owen*, Mid. Eo.   Part I, p. 31, no plate.

Crocodile de Sheppy (?) *Cuvier*.   Part II, pp. 29 and 31.

Crocodilus champsoides, *Owen*, Lond. Cl.   Part II, p. 31, tab. ii, fig. 2?, and tab. iii.

   „   Hastingsiæ, *Owen*, Mid. Eo.   Part II, pp. 37 and 44, tabs. vi, vii, viii, fig. 1, ix and xii, figs. 2, 5.

   „   *Spenceri, Buckland*.   Part II, p. 29.

   „   toliapicus, *Owen*, Lond. Cl.   Part II, p. 29, tab. ii, fig. 1, tab. v, figs. 1, 2, 3, 5, 6, and 12, tabs. ii, ii A, fig. 1, and tab. iv.

INDEX.

Emys bicarinata, Lond. Cl., *Bell.*   Part I, p. 73, tabs. xxv, xxvi.

„   Comptoni, Lond. Cl., *Bell.*   Part I, p. 71, tab. xx.

„   Conybeari, *Owen*, Lond. Cl.   Part I, Sup. I, p. 77, tabs. xxviii A and xxviii B.

„   crassus, *Owen*, Mid. Eo.   Part I, p. 76, tab. xxvii.

„   Delabechii, Lond. Cl., *Bell.*   Part I, p. 74, tab. xxviii.

„   *de Sheppy, Cuv.* (?)   Part I, p. 67, tab. xxiv.

„   „   *H. v. Meyer.* (?)   Part I, p. 10.

„   lævis, *Bell*, Lond. Cl.   Part I, p. 70, tab. xxii.

„   *Parkinsoni, J. E. Gray.*   Part I, p. 10.

„   testudiniformis, *Owen.*   Part I, p. 67, tab. xxiv.

Gavialis Dixoni, *Owen*, Mid. Eo.   Part II, p. 42, tab. x.

Palæophis (?) longus, *Owen*, Low. Eo.   Part III, p. 66, tab. xiv, figs. 35—37, 45, 46.

„   porcatus, *Owen*, Mid. Eo.   Part III, p. 61, tab. xiv, figs. 13—15, 18, 20, 21.

„   toliapicus, *Owen*, Lond. Cl.   Part III, p. 63, tabs. xv and xvi.

„   typhæus, *Owen*, Mid. Eo.   Part III, p. 56, tab. xii, figs. 5—8, tab. xiv, figs. 1—3, 7—9, 16, 17, 26, 27, 28 (figs. 6, 10—12) (?).

Paleryx depressus, *Owen*, Mid. Eo.   Part III, p. 68, tab. xiii, figs. 37, 38.

„   rhombifer, *Owen*, Mid. Eo.   Part III, p. 67, tab. xiii, figs. 29—32.

Platemys Bowerbankii, *Owen*, Lond. Cl.   Part I, p. 66, tab. xxiii.   Part II, Sup. I, p. 1, tab. xxix, figs. 1, 2.

„   Bullockii, *Owen*, Lond. Cl.   Part I, p. 62, tab. xxi.

Testudo plana, König.   Part I, p. 27.

Trionyx Barbaræ, *Owen*, Mid. Eo.   Part I, p. 50, tab. xvi A.

„   circumsulcatus, *Owen*, Mid. Eo.   Part I, p. 59, tab. xix B, figs. 1—3.

„   Henrici, *Owen*, Mid Eo.   Part I, p. 46, tab. xvi.

„   incrassatus, *Owen*, Up. Eo.   Part I, p. 51, tabs. xvii, xviii, and xix.

„   marginatus, *Owen*, Mid. Eo.   Part I, p. 55, tab. xix* and tab. xix B, figs. 4—6.

„   planus, *Owen*, Mid. Eo.   Part I, p. 58, tabs. xix C and xix D, fig. 6.

„   pustulatus, *Owen*, Lond. Cl.   Part I, p. 60, tab. xix B, figs. 7—9.

„   rivosus, *Owen*, Mid. Eo.   Part I, p. 56, tab. xviii A.

„   Sp., *Owen.*   Part I, Mid. Eo., tab. xix D, figs. 3—5.

„   Sp., *Owen.*   Part I, Mid. Eo., p. 61, tab. xix D, fig. 7.

MONOGRAPH

ON

THE FOSSIL REPTILIA

OF THE

LONDON CLAY, AND OF THE BRACKLESHAM AND OTHER TERTIARY BEDS.

PART I.

PAGES 1—76; PLATES I—XXVIII, and VIIIA, XA, XIIIA, XIIIB, XVIA, XVIIIA, XIX*, XIXB, XIXC, XIXD.

CHELONIA (CHELONE, &c.).

BY

PROFESSOR OWEN, D.C.L., F.R.S., F.L.S., F.G.S., &c.

AND

PROFESSOR BELL, F.R.S., F.L.S., F.G.S., &c.

Issued in the Volume for the Year 1848.

LONDON:

PRINTED FOR THE PALÆONTOGRAPHICAL SOCIETY.

1849.

MONOGRAPH

ON THE

FOSSIL REPTILIA OF THE LONDON CLAY.

ORDER.—*CHELONIA.*

Family—MARINA.

Genus—CHELONE.

THE majority of the Fossil Chelonians of the Eocene tertiary deposits, defined or described in my ' Report on British Fossil Reptiles,' belonged to the marine division of the order, and to the genus *Chelone;* and as the species of this genus depart least from the ordinary reptilian type in the modification of the bones of the trunk, composing the characteristic thoracic-abdominal case of the order, I propose to commence with them those descriptions of the Chelonian reptiles which fall to my share of the present Monograph.

In order to facilitate the comprehension of the descriptions and figures of the fossil Chelonians, a brief notice is premised of the composition and homologies of the carapace and plastron, or roof and floor, of that singular portable abode, with which the reptiles of the present order have been endowed in compensation for their inferior powers of locomotion or other modes of escape or defence.

In the marine species of the Chelonian order, of which the *Chelone mydas* may be regarded as the type, the ossification of the carapace and plastron is less complete, and the whole skeleton is lighter than in those species that live and move on dry land: but the head is proportionally larger—a character common to aquatic animals,— and being incapable of retraction within the carapace, ossification extends in the direction of the fascia, covering the temporal muscles, and forms a second bony covering of the cranial cavity: it is interesting to observe, however, that this accessory defence is not formed by the intercalation of any new bones, but is due to exogenous growth from the frontals (11), parietal (7), postfrontals (12), and mastoids (8, see T. I, T. III, T. XV).

The bony carapace is composed externally of a series of median and symmetrical pieces (fig. 1, ch, $s1$—$s11$, py), and of two series of unsymmetrical pieces ($pl1$—8, $m1$—12) on each side. The median pieces have been regarded as lateral expansions of the summits of the upper vertebral (neural) spines,* the median lateral pieces as similar

* Cuvier, Lecons d'Anatomie Comparée, tom. i (1799), p. 212.

1

developments of the vertebral ribs (pleurapophyses),* and the marginal pieces as the homologues of the sternal ribs (hæmapophyses).†

I must refer the reader to my Memoir, communicated to the Royal Society, for the facts and arguments which have led me to regard these pieces, as dermal ossifications, homologous with those that support the nuchal and dorsal epidermal scutes in the crocodile. Most of the bony pieces of the carapace are, however, directly continuous, and connate,‡ with the obvious elements of the vertebræ, which have been supposed exclusively to form them by their unusual development; the median pieces have accordingly been called "vertebral plates," and the medio-lateral pieces "costal plates." I retain the latter name, although with the understanding and conviction that they are essentially or homologically distinct parts from the vertebral ribs or pleurapophyses with which they are connate and more or less blended. But, with regard to the term "vertebral" plate, since the ribs (*costæ*) are as essentially elements of the vertebra as the spinous processes themselves, I have been in the habit, in my Lectures, of indicating the median series by the term "neural plates," which term has the further advantage of removing any ambiguity from the descriptions that might arise from their being mistaken for the superincumbent epidermal shields, which are likewise called "vertebral plates" in some English works.§ The term "marginal" is retained for the osseous plates forming the periphery of the carapace; but the median and symmetrical ones, which seem also to begin and end the "neural" series, are specified, the one by the term "nuchal plate," the other by that of "pygal plate." The "neural plates" are numbered as in the classical Monograph of Bojanus.‖

In the subjoined woodcut of the carapace of the loggerhead turtle (*Chelone caouanna*) (fig. 1), *ch* is the *nuchal plate*; $s1$ to $s11$ the *neural plates*; $pl1$ to $pl8$ the *costal plates*; and $m1$ to $m12$ the marginal plates. The carapace is impressed by the superimposed epidermal scutes or shields, which consist of a median series, called "*vertebral scutes*" $v1$ to $v5$;

---

* Ibid. p. 211. Rathké has recently supported this determination by arguments drawn from the mode of development of the carapace. See 'Annales des Sciences Naturelles,' Mars, 1846; and 'Ueber die Entwickelung der Schildkröten,' 4to, 1848, where he says, p. 105 :—" Ausser den Rippen und den horizontal liegenden Tafeln, zu welchen sich die Dornfortsätze des zweiten und der sechs folgenden Rückenwirbel ausbilden, dienen bei den erwachsenen Schildkröten zur Zusammensetzung des Rückenschildes noch eine oder mehrere Knochenplatten," viz. the "marginal plates." I have shown how Rathké was deceived by over-estimating the character of connation, in my 'Observations on the Development of the Carapace and Plastron of the Chelonians,' which conduct to a different conclusion to that at which Cuvier and Rathké have arrived. (Philosoph. Transactions, 1849.)

† Geoffroy, Annales du Muséum, tom. xiv (1809), p. 7.

‡ This term is used in the definite sense explained in my work on the 'Archetype of the Vertebrate Skeleton' (8vo, V. Voorst, p. 49), as signifying those essentially different parts which are not physically distinct at any stage of development; and in contradistinction to the term "confluent," which applies to those united parts which were originally distinct.

§ See Griffiths's translation of Cuvier, vol. ix, Synopsis of Reptilia, p. 6—"fifth vertebral plates prominent."

‖ Anatome Testudinis Europææ, fol. 1821, tab. iii and iv.

and of a lateral series of "*costal scutes;*" there is also a peripheral series of "*marginal scutes*" corresponding with and impressing the marginal plates. The nuchal plate (*ch*) is remarkable for its breadth in all Chelonia, and usually sends down a ridge from the middle line of its under surface, which is attached by ligament to the summit of the neural arch of the first dorsal vertebra. The first true neural plate, $s1$, is much narrower, and is connate with the summit of the neural spine of the second dorsal vertebra; the succeeding vertebral neural plates, $s2$—$s8$, have the same relations with the succeeding neural spines, but the ninth, tenth, and eleventh, like the nuchal (*ch*) and pygal (*py*), plates are independent ossifications in the substance of the derm. The costal pieces of the carapace are supra-additions to

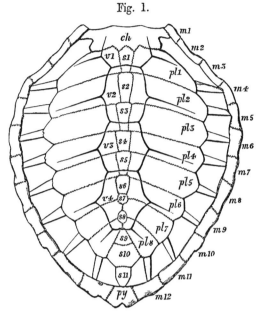

Fig. 1.

Carapace of the Loggerhead Turtle (*Chelone caouanna*).

eight pairs of pleurapophyses or vertebral ribs, those, viz. of the second to the ninth dorsal vertebræ inclusive. The slender or normal portions of the ribs project freely

for some distance beyond the expanded and connate portions ("costal plates" of the carapace), along the under surface of which the rib may be traced, of its ordinary breadth, to the neck and head, which liberates itself from the costal plate to articulate to the interspace of the two contiguous vertebral bodies, (centrums), to the posterior of which such rib properly belongs.

The woodcut (fig. 2) illustrates this structure: *ch* shows the inner side of the nuchal plate; *c1* is the first rib, articulated to the fore part of the body of the first dorsal vertebræ; *pl1* is the first rib of the carapace (the second rib of the dorsal series), connate with the first costal plate; *pl2* to *pl8*, are the succeeding ribs and costal plates of the carapace. The heads of the ribs articulate to

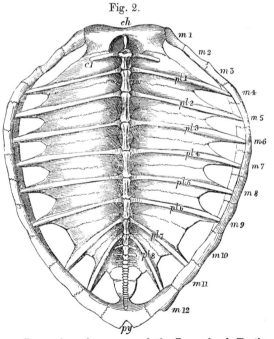

Fig. 2.

Inner view of carapace of the Loggerhead Turtle (*Chelone caouanna*).

the interspaces between their own vertebral body, and that of the preceding vertebra. The tenth vertebra supports a short pair of ribs in *Chelone* and in *Emys*, but not in *Trionyx;* and this vertebra is commonly reckoned as a "lumbar" one. The eleventh and twelfth vertebræ have short and thick ribs, which abut against the iliac bones, and they are regarded as forming the sacrum. The remaining vertebræ belong to the tail, and are "caudal." The costal plates articulate with each other, and with the neural plates by fine dentated sutures. The free extremities of the ribs are implanted into sockets of those marginal plates which are opposite to them. The 1st, 2d, 3d, and 10th, are not so articulated in the loggerhead turtle. But all the marginal plates articulate with each other, and with the nuchal (*ch*) and pygal (*py*) plates by sutures.

The osseous basis of the plastron consists of nine pieces, one single and symmetrical, the rest in pairs.

The median piece, *s*, is the *entosternal;* the anterior pair, *es*, is the *episternal;* the second pair, *hs*, the *hyosternal;* the third pair, *ps*, the *hyposternal;* and the posterior pair, *xs*, the *xiphisternal.*

With regard to the nature or homologies of these bones, three views have been

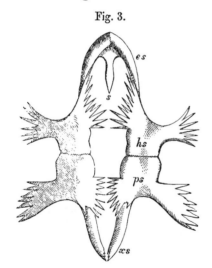

Fig. 3.

Bones of the plastron of the Loggerhead Turtle (*Chelone caouanna*).

taken. The one generally adopted, on the authority of Cuvier, Bojanus, and Geoffroy St. Hilaire, is, that the nine bones of the plastron are subdivisions of a vastly expanded sternum, or breast-bone; the second view is, that these subdivisions of the sternum are enlarged by combination with ossifications of the integument;* and the third view, in which Rathké stands alone, is, that they are exclusively dermal bones, and have no homologues in the endoskeleton of other vertebrata.†

Since this opinion is given as the result of that celebrated embryologist's observations on the development of the Chelonian reptiles, I have tested it by a series of similar researches on the embryos and young of the *Chelone mydas* and *Testudo indica*, and have been led by them to conclusions distinct from any of the three theories above cited.

The sternum, like the carapace, is, without doubt, a compound of connate, endoskeletal and exoskeletal pieces; but the endoskeletal parts are not exclusively the homologues of the sternum. For the details of the observations, and the special arguments on which these conclusions are founded, I must refer to my paper in the 'Transactions of the Royal Society,' 1849; the homologies of the endoskeletal parts of the plastron will require a brief illustration here from comparative anatomy.

* Peters, Observationes ad Anatomiam Cheloniorum, 1838.
† Ueber die Entwickelung der Schildkröten, 4to, 1848, p. 122.

Geoffroy St. Hilaire, whose views are generally adopted, was guided in his determination of the parts of the plastron by the analogy of the skeleton of the bird: which analogy may be illustrated by the subjoined diagrams of corresponding segments of the thorax of a bird (fig. 4) and of a tortoise (fig. 5). In both figures c is the centrum or vertebral body; ns the neural arch and spine; compressed in the bird, depressed and laterally expanded, according to Geoffroy, in the tortoise; pl the pleurapophysis, or vertebral rib, expanded in the tortoise, and with its broad tubercle articulating with the expanded spine; h, h' in fig. 5, answers to h in fig. 4, and is the hæmapophysis (sternal rib, or ossified cartilage of the rib); h, hs in fig. 5, is hs in fig. 4, i. e. exclusively a sternum, with the entosternal piece, hs', developed horizontally in the tortoise, and vertically in the bird. The *primâ facie* simplicity of this view has imposed upon most comparative anatomists: and yet there are other vertebrate animals more nearly allied to the *Chelonia* than birds, and with which, therefore, comparison should have been instituted before general consent was yielded to the Geoffroyan hypothesis.

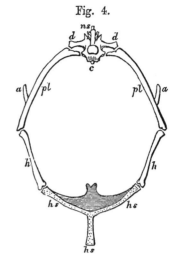

Fig. 4.

Thoracic segment of the skeleton of a Bird.

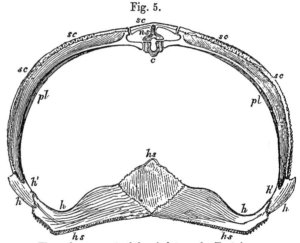

Fig. 5.

Thoracic segment of the skeleton of a Tortoise.

If, e. g. we take the segment of a crocodile's skeleton (fig. 6) corresponding with that of the tortoise (fig. 5), the comparison will yield the following interpretation: in both figures c is the centrum: ns the neural arch and spine, with d the diapophysis; sc a median dermal bony plate (connate with ns in the tortoise); pl the pleurapophysis; sc sc lateral dermal bony plates (connate with pl in the tortoise); h, h' in fig. 5, answers to h' in fig. 6, an intercalated, semi-ossified piece between pl and h in the crocodile; h, hs in fig. 5, answers to h, the hæmapophysis in the crocodile; and hs in fig. 5, exclusively represents hs, the sternum in the crocodile.

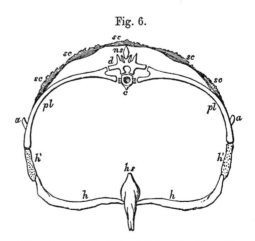

Fig. 6.

Thoracic segment of the skeleton of a Crocodile.

Such a comparison, in my opinion, guides us to a truer view of the homologies of the thoracic-abdominal bony case of the Chelonians, especially with regard to the lateral or parial pieces of the plastron, than the comparison exclusively relied on by Geoffroy St. Hilaire. The *Plesiosaurus*, by its long and flexible neck, small head, expanded coracoid and pubis, and flattened bones of the paddles, comes much nearer to the turtle than the crocodile does; and its abdominal ribs, or hæmapophyses, are more developed than in the crocodiles; a comparison of the ventral surface of the skeleton, such as that figured by Dr. Buckland, in his 'Bridgewater Treatise,' vol. ii, pl. 18, fig. 3, will show how clearly those abdominal ribs would correspond with the hyosternals and hyposternals of the turtle, if they had coalesced together at their middle parts, leaving their outer and inner extremities free.

With regard to the marginal pieces m_1—m_{12}, figs. 1 and 2, although the comparisons illustrated by figs. 4, 5, 6, show that they answer rather to the intercalated piece h' in the crocodile than to the entire sternal rib h in the bird; yet the phenomena of their development demonstrate that they are exclusively bones of the dermal skeleton, retaining their freedom from anchylosis with the endoskeletal elements, like the nuchal, pygal, and last three neural plates (ch, py, s_9, s_{10}, and s_{11}, fig. 1). This insight into their true nature teaches why they do not correspond in number with the vertebral ribs or pleurapophyses (pl_1—pl_8, fig. 2). In the loggerhead turtle, for example, the first three and the tenth (m_1, m_2, m_3, and m_{10}) have no corresponding pleurapophyses articulating with them; and if even c_1 be supposed to correspond to m_3, there are no rudiments of ribs answering to m_1 and m_2. The marginal plates are not constant in number; the *Chelone mydas* has two less than the *Chelone caouanna* has. Some species of *Trionyx* (*Cryptopus*, Dum. and Bibron) have a greater number, but of smaller and less regular size, confined to the posterior part of the limb of the carapace; in other species of *Trionyx* (*Gymnopus*, Dum. and Bibron), and in *Sphargis*, the marginal part of the carapace retains its embryonic condition in all *Chelonia*, as a stratum of cartilaginous cells in the substance of the derm, forming the thickened, flexible border of the carapace.

The rudiments of the hyosternals and hyposternals have originally the form of sternal or abdominal ribs; extend transversely, and rise at their outer extremities to join those of the first and sixth pair of vertebral ribs, completing the hæmal, or inferior vertebral arch, without the interposition of any of the marginal pieces, which are merely applied to the outer sides of the hæmapophysis or sternal ribs. The expansion of the parts of the plastron, especially in the fresh-water and land tortoises, is due chiefly to the ossification of a layer of cartilage-cells in the substance of the derm, which ossified plates are connate with the more internal elements of the plastron, representing the sternum and sternal ribs. In the following descriptions of the fossil *Chelonia*, the terms 'entosternal, episternal, hyosternal, hyposternal,' and 'xiphisternal,' will be used as absolute designations of the combined endoskeletal and exoskeletal bones of the plastron, without implying assent to the hypothesis that first suggested those names to Geoffroy St. Hilaire.

The scapular and pelvic arches, and the bones of the extremities of the *Chelonia*, are described and figured in the 'Ossemens Fossiles' of Cuvier;* where, also, the figures of the modifications of the carapace and plastron, in the fresh-water and land tortoises, will suffice for the purpose of ulterior comparisons with the fossils described in the present work, if they be understood according to the homologies above discussed, and which are illustrated by the figures 1 and 2 of the carapace, and fig. 3 of the plastron of the *Chelone caouanna*.

With regard to the more immediate subjects of the present Monograph, it must be admitted that the important generalizations of Cuvier and Dr. Buckland† have been confirmed, but not materially extended, by subsequent observations on the remains of reptiles of the Chelonian order. Cuvier, after admitting that his results in regard to the tortoises were not so precise as those relating to the crocodiles, sums up his chapter on the fossil *Chelonia* in the following words : " Toutefois nous avons pu nous assurer que les tortues sont aussi anciennes dans le monde que les crocodiles ; qu'elles les accompagnent généralement, et que le plus grand nombre de leurs débris appartenant à des sous-genres dont les espèces sont propres aux eaux douces ou à la terre ferme, elles confirment les conjectures que les os de crocodiles avoient fait naître sur l'existence d'iles ou de continens nourissant des reptiles, avant qu'il y ait eu des quadrupèdes vivipares, ou du moins avant qu'ils aient été assez nombreux pour laisser une quantité de débris comparable à ceux des reptiles."‡

Dr. Buckland also states, in general but precise terms, that " the Chelonian reptiles came into existence nearly at the same time with the order of *Saurians,* and have continued coextensively with them through the secondary and tertiary formations unto the present time. Their fossil remains present also the same threefold divisions that exist among modern *Chelonia* into groups, respectively adapted to live on land, in fresh water, or the sea."§

The remains of sea turtles (*Chelone*) have been recognised in the Muschelkalk, the Wealden, the lower cretaceous formation at Glaris, and the upper chalk-beds at Maestricht. Figures of Chelonites, as that in the Frontispiece to Woodward's 'Synoptical Table of British Organic Remains,' and in König's 'Icones Sectiles' (pl. xviii, fig. 232, *a* and *b*), have been published ; but no true marine Chelonian, from Eocene strata, had been scientifically determined prior to the communication of my Paper on that subject to the Geological Society of London.‖ All the Chelonites from Sheppey, described and figured in the last edition of Cuvier's 'Ossemens Fossiles,' for

---

* Tom. v, pt. 2, pl. xii and xiii.

† Bridgewater Treatise (1836), p. 256.

‡ Ossemens Fossiles, 4to, tom. v, pt. ii, p. 249.

§ Bridgewater Treatise, p. 256.

‖ Proceedings of the Geological Society of London, vol. iii, pt. ii, p. 570, December 1, 1841.

example, are referred to the fresh-water genus *Emys;* and the statement in the earlier edition of the 'Ossemens Fossiles,' that the greater part of the remains of Chelonian reptiles belong to the fresh-water or terrestrial genera, is repeated.

The aim of the Memoir, communicated to the Geological Society in December, 1841, was to show that the conclusion deduced by Cuvier, from an imperfect carapace from Sheppey, which might probably have belonged to a species of *Emys,* had been unduly extended to other Chelonites, which undoubtedly belonged to the marine genus *Chelone;* and that this genus was represented, in the Eocene strata, by at least six species; the remains of five of which were from the London Clay at Sheppey, and those of a sixth were tolerably abundant in the cliffs near Harwich.

In the carapace of the fossil Chelonian from Sheppey, communicated by Mr. Crowe, of Faversham, to Cuvier, and figured in the ' Ossemens Fossiles (tom. v, part 2, pl. xv, fig. 12), the author of that great work conceived that all the characters of the genus *Emys* were perfectly recognisable.

He points out the proportions of the neural plates, which are as long as they are large; and in the figure they are represented of nearly a quadrate form, and not rhomboidal.

The fifth neural plate in the fragment figured (probably the eighth) is separated from the sixth (ninth) by a point, which is made by the mesial ends of the fifth (probably the seventh) pair of costal plates; a structure which Cuvier says slightly recalls what he had observed in the Jura *Emys* of Soleure.*

But Cuvier admits that the neural plates (*plaques vertébrales*) are narrower than those of existing *Emydes ;* and that the equal breadth of the ribs is a character common to the *Chelones* with the *Emydes.*

Now, in reference to the carapace figured by Cuvier, it is to be observed, that the margins are wanting; and that the broad conjoined portions of the costal plates are not longer than they might have been, had the fossil belonged to a turtle (*Chelone*); and, consequently, that there is no proof that they were united together by suture throughout their whole extent, as in the *Emydes ;* but that they might have terminated in narrow tooth-like processes, as in the *Chelones.*

The narrowness of the neural plates is a character which, with their smoothness, undoubtedly approximates the fossil to the *Chelones;* and, without intending to affirm that the fossil in question does not belong to the family *Emydidæ,* which unquestionably existed at the time of the deposition of the Sheppey clay, its determination appears to me to be much less decisive than might be inferred from the remarks in the ' Ossemens Fossiles.'

* Tom. cit., p. 234. This structure is not, however, peculiar to the genus *Emys ;* in the carapace of the *Chelone caouanna,* in the Museum of the Royal College of Surgeons, the seventh neural plate is separated from the eighth by the junction of the expanded extremity of the seventh rib on one side with that of the opposite rib, and the eighth neural plate from the ninth by the same modification of the eighth pair of ribs. A similar modification may also be seen in the carapace of the *Trionyx Henrici,* T. XVI.

Mr. Parkinson describes the plastron of a Sheppey Chelonite,* in which the hyosternal and hyposternal pieces are not united, but leave a vacancy in the middle, which he conjectured may have been filled up by membrane. This specimen must have belonged to a specimen at least four inches in length, exclusive of the head and neck. But Cuvier supposes that it may, nevertheless, have belonged to an *Emys;* and that the vacancy of the bony sternum merely indicated the nonage of the individual.†

The grounds on which Cuvier refers to the genus *Emys,* the imperfect and dislocated carapace and plastron of M. Bourdet's Sheppey Chelonite,‡ are not detailed; but it is evident that the hyposternals in that specimen are in contact at the posterior moiety of their median margins only; and that the margins recede anteriorly, leaving a median interspace; which, as the plastron is nearly a foot in length, can hardly be attributed to the immature state of the individual. And if, as Cuvier supposes, this specimen belongs to the same species as those in the collections of Messrs. Crowe and Parkinson, the same objection to their belonging to a fresh-water tortoise holds good, as to the one figured by M. Bourdet.

The question of the reference of these Eocene fossils to the fresh- or sea-water families of the Chelonian order, seems to me to admit of the safest determination by examining the crania of the Sheppey Chelonites; since the differences in the extent to which the temporal fossæ are protected by bone, and in the proportions in which the bones enter into the formation of that covering, are strongly marked in the genera *Emys* and *Chelone.*

But here Cuvier appears to have been unusually biassed in favour of the Emydian nature of the Sheppey fossils; for in reference to the cranium, figured by Mr. Parkinson, the affinities of which to the turtle's skull will be presently pointed out, Cuvier observes : "elle est probablement aussi d'une Emyde, bien qu'elle participe des caractères de Tortues de Mer, par la manière dont le parietal recouvre sa tempe ; mais nous avons vu que *l'Emys expansa* diffère très peu de Tortues de Mer à cet égard, et la partie antérieure de la tête fossile ressemble d'avantage à celle d'une Emyde qu' à celle d'une Chelonée, surtout par le peu de largeur de l'intervalle des yeux."§

Now the most striking difference between the temporal bony vault of the *Emys expansa* and that of any known species of *Chelone,* is seen in the diminutive size of the post-frontals in this exceptional case among the *Emydes,* as contrasted with their large size and actual extension over the temporal fossæ in the *Chelones:*—and this difference is accompanied by a proportional diminution in the breadth of the parietals in the true marine turtles.

* Organic Remains, vol. iii, p. 268, pl. xviii, fig. 2.
† Ossemens Fossiles, tom. v, pt. ii, p: 235.
‡ Tom. cit., pl. xv, figs. 14-15.
§ Tom. cit., p. 235.

2

But the figure in Parkinson's work gives clearly the latter character; whence also we may infer that it agreed more with the *Chelones* also in the size of the postfrontals; although the anatomy of the skull is too obscurely delineated to demonstrate this fact.

The following important affinities are, however, unquestionably indicated in Parkinson's figure:—*first*, the large size of the orbits, which are nearly six times greater than those of the *Emys expansa; secondly*, their more posterior and lateral position; and *thirdly*, the greater breadth of the interorbital space: in all which characters the Sheppey fossil closely resembles the true *Chelones*, and differs from the only known species of *Emys (Podocnemys) expansa*, in which the temporal openings are protected by a bony roof.

That fresh-water tortoises have left their bony cuirasses in the Sheppey clay, will be subsequently shown; but the evidence of the genus *Emys*, adduced by Cuvier, is incompetent to prove their existence; and, it may be affirmed, that of the fossils cited by the founder of Palæontology, some, with great probability, and others with certainty, are referable to the marine genus, *Chelone*.

Without further discussing the question as regards these evidences, I shall proceed to describe the specimens from Sheppey which I have myself had the opportunity of examining; and shall commence with those which belong undoubtedly to the marine family.

CHELONE BREVICEPS.   *Owen*. Tabulæ I and II.
Proceedings of the Geological Society, December 1, 1841; Report on British Fossil
Reptiles, Trans. British Association, 1841, p. 178.
Syn. EMYS PARKINSONII.   *J. E. Gray*.
— DE SHEPPEY.   *H. v. Meyer* (?).
CHELONE ANTIQUA.   *Kœnig* (?).

The first of the Chelonites, which led me to the recognition of this species, was a nearly perfect cranium from Sheppey (Tab. I, figs. 1—4), wanting only the occipital spine, and presenting a strong and uninterrupted roof, extended posteriorly from the parietal spine on each side (7, 7), over the temporal openings to the mastoids (8, 8); and formed anteriorly by a great development of the posterior frontals (22).

This unequivocal testimony of the marine genus of the fossil, is accompanied by similar evidence afforded by the large size and lateral aspect of the orbits, the posterior boundary of which extends beyond the anterior margin of the parietals; and by the absence of the deep emargination which separates the superior maxillary from the tympanic bone in the fresh-water tortoises, and especially in the *Podocnemys expansa*.

In general form, the skull of the present species of Sheppey *Chelone* resembles that of the *Chelone mydas*, BRONGN.: but it is relatively broader; the prefrontals (14) are less sloping, and the anterior part of the head is more vertically truncate. The orbits are relatively larger, and extend nearer to the tympanic cavity. The frontals (11) enter into the formation of the orbits in rather a larger proportion than in *Chelone mydas*. In the *Chelonè caouanna** they are wholly excluded from the orbits.

The trefoil shape of the occipital tubercle is well marked (fig. 4) ; the depression in the basioccipital, bounded by the angular pterygoid ridges, is as deep as in most true turtles (fig. 3) ; the lateral borders of the expanded parietals are united by a straight suture along a great proportion of their extent to the large postfrontals.

These proportions are reversed in the *Podocnemys expansa*, in which the similarly expanded plate of the parietals is chiefly united laterally with the squamosal and tympanic bones. In other fresh-water tortoises the parietal plate in question does not exist.

The same evidence of the affinity of the Sheppey Chelonite in question to the marine turtles, is afforded by the base of the skull (fig. 3) ; the basioccipital (1) is deeply excavated; the processes of the pterygoids (24), which extend to the tympanic pedicles, are hollowed out lengthwise : the palatal processes of the maxillary and palatine bones are continued backwards to the extent which characterises the existing *Chelones;* and the posterior or internal opening of the nasal passages, is, in a proportional degree, carried further back in the mouth. The lower opening of the zygomatic spaces is wider in the present Sheppey Chelonite, than in *Podocnemys expansa.*

The external surface of the cranial bones in the fossil is roughened by small irregular ridges, depressions, and vascular foramina, which give it a wrinkled or shagreen-like character.

The following are dimensions of the specimen described :

	Inches.	Lines.
Length of cranium from the occipital condyle	2	9
Breadth of cranium across the malars (26)	2	7
Antero-posterior diameter of orbit	1	0

The lower jaw, which is preserved in the present fossil, likewise exhibits two characters of the marine turtles ; the dentary piece (32), *e. g.* forms a larger proportion of the lower jaw than in the land or fresh-water tortoises. The joint of the rami is completely obliterated at the symphysis, which is not longer or larger than in *Chelone mydas*.

The species represented by this fossil, which is preserved in the British Museum, and by a very similar one in the Hunterian Collection, is selected for the first of the Eocene Chelonians to be described in the present Monograph, because it is one of the few with which the characters of the carapace and plastron can with certainty be associated with those of the cranium.

* Ossem. Fossiles, tom. v, pt. ii, pl. xi, fig. 2.

In the rich collection of Sheppey fossils, belonging to J. S. Bowerbank, Esq. F.R.S. there is a beautiful Chelonite (Tab. II, figs. 1, 2) including the carapace, plastron, and the cranium, which is bent down upon the fore part of the plastron; and which, though mutilated, displays sufficient characters to establish its specific identity with the skull of the *Chelone breviceps* just described. Both the carapace and plastron present the same finely rugous surface externally as the cranium; in which character we may perceive a slight indication of affinity with the genus *Trionyx*.

The carapace (T. II, fig. 1) is long, narrow, ovate, widest at its anterior half, and tapering towards a point posteriorly; it is not regularly *convex*, but slopes away, like the roof of a house, from the median line (fig. 3), resembling, in this respect, and its general depression, the carapace of the turtle *Chelone mydas*. There are preserved the nuchal plate (fig. 1, *ch*) with ten of the neural plates ($n1$—$n10$), only the eleventh and pygal plates being wanting. The eight pairs of costal plates ($pl1$—$pl8$) are also present, with sufficient of the narrower tooth-like extremities of the six anterior pairs of ribs, to determine the marine character of the fossil, which is indicated by its general form.*

The nuchal plate (fig. 6, *ch*) is of a transversely oblong form, with the anterior margin gently concave. Its antero-posterior diameter, or length, is ten lines; its transverse diameter, or breadth, is two inches. The lateral margins are bounded by two lines meeting at a slight angle; to the anterior one, the first of the marginal plates, $m1$, is attached; the posterior line bounds part of the vacant interspace between the first costal plate ($pl1$), and the anterior marginal plate. The presence of this plate would prove for the genus *Chelone* as against *Trionyx*, were the characters of the cranium, the impressions of the vertebral scutes, and the sternum wanting. The nuchal plate in the *Emydes* is hexagonal, and nearly as long as it is broad.

The Chelonite from the tertiary beds near Brussels, figured by Cuvier,† has the nuchal plate of nearly the same form as the present specimen from Sheppey.

The neural plates in the *Chelone breviceps* are as narrow as in the *Chelones* generally; and as in the Brussels Chelonite above cited.

The first neural plate ($s1$, fig. 1) is four-sided; the rest, to the eighth ($s8$), are hexagons of a more regular figure than in the existing *Chelones*, and are articulated to more equal shares of the contiguous alternate costal plates ($pl1$—$pl8$).

The first costal plate ($pl1$) is directed more outwards, does not incline backwards, as in recent *Chelones*, and its anterior angle is less truncated than in them. (See fig. 1, p. 3.)

The length of the second costal plate ($pl2$) is one inch, nine lines; more than half of the narrow terminal extremity of the connate rib is preserved; the proportions of

---

* In an *Emys* with a carapace seven inches in length, the corresponding extremities of the ribs would have been united together by the laterally-extended ossification.

† Ossemens Fossiles, tom. v, pt. 2, pl. xv, fig. 16.

the remaining costal plates correspond with those of the *Chelone mydas*, and *Chel. caouanna*.

The last pair of costal plates (*pl8*) articulates with the eighth, ninth, and tenth neural plates, but does not overlap or supersede any of them.

Not any of the costal plates articulate with those of the opposite side, so as to interrupt the series of vertebral plates, as in the carapace of the *Chelone caouanna* (fig. 1, p. 3), as in Mr. Crowe's Sheppey Chelonite, figured by Cuvier (tom. cit. pl. xv, fig. 12); and as is shown in the view of the concave surface of the Brussels species (tom. cit. pl. xv, fig. 16).

The ninth neural plate (fig. 1, *s9*) is the narrowest, as in the *Chelones*, and as in the Brussels Chelonite, figured by Cuvier, in loc. cit. pl. xiii, fig. 8, instead of being suddenly expanded, as in most *Emydes*.

The tenth neural plate (*s10*) expands to a breadth equal with its length; the eleventh and pygal plates, as already observed, are wanting in the fossil.

The vertebral or median ends of the costal plates present a modification of form, corresponding with that of the interspaces of the neural plates to which they are articulated. Only the first pair (*pl1*) present that form which characterises all but the last pair in the existing *Chelones*, and in the Brussels Chelonite; viz., a straight line with the posterior angle cut off; the rest being terminated by two nearly equal oblique lines, meeting at an open angle, as shown in Tab. II, fig. 1, *pl2—pl7*.

This character would serve to distinguish the *Chelone breviceps*, if only a portion of the carapace, including the vertebral extremity of a rib, were preserved. The free extremities of the ribs are thicker in proportion to the costal plates, than in the *Chelone caouanna*, or the *Chel. mydas*; and more resemble, in this respect, those of the *Chel. imbricata*, the species characterised by the size and beauty of the horny scutes, commonly called " tortoise-shell."

More or less complete impressions of the five horny vertebral scutes (*v1—v5*), and of four costal scutes on each side of the vertebral ones, show the forms and proportions of these characteristic parts, and especially of the median series, notwithstanding they were among the soluble and perishable elements of this ancient turtle of the Thames.

The hexagonal vertebral scutes are characterised by the near equality of their sides, and the angle of about 100°, at which the two outer sides meet.

The anterior border of the first vertebral scute, v^1, has crossed and impressed the nuchal plate, *ch*, near its anterior border; this scute has covered the rest of the nuchal plate, and more than half of the first neural plate. The second vertebral scute, v^2, includes the rest of the first neural plate, the whole of the second, and almost the whole of the third neural plate. The third vertebral scute, v^3, includes the hind border of the third neural plate, with the whole of the fourth and fifth neural plates. The fourth vertebral scute includes the sixth and seventh, and very nearly the whole of the eighth neural plates, and the outer angles of this scute terminate over the suture between the sixth and seventh costal plates.

The plastron of the *Chelone breviceps* (Tab. II, fig. 2), although more ossified than in existing *Chelones*, yet presents all the essential characters of that genus. There is a central vacuity left between the hyosternals (*hs*) and hyposternals (*ps*); but these bones differ from those of the young *Emys* in the long pointed processes which radiate from the two anterior angles of the hyosternals (*hs*), and the two posterior angles of the hyposternals (*ps*).

The xiphisternals (*xs*) have the slender elongated form, and oblique union by reciprocal gomphosis with the hyposternals (*hs*), which is characteristic of the genus *Chelone*.

The posterior extremity of the right episternal (*es*) presents the equally characteristic, slender pointed form.

With these proofs of the modification of the plastron of the present fossil according to the peculiar type of the marine *Chelones*, there is evidence, however, that it differs from the known existing species in the more extensive ossification of the component pieces; thus the pointed rays of bone extend from a greater proportion of the margins of the hyosternals and hyposternals; and the intervening margins do not present the straight line at right angles to the radiated processes.

In the *Chelone mydas*, and *Chel. caouanna* (fig. 3, p. 4), for example, one half of the external margin of the hyosternal and hyposternal, where they are contiguous, are straight, and intervene between the radiated processes, which are developed from the remaining halves, while in the *Chelone breviceps*, about a sixth part only of the corresponding external margins are similarly free, and there form the bottom, not of an angular, but a semicircular interspace.

The radiated processes from the inner margins of the hyosternals and hyposternals, are characterised in the *Chelone breviceps* by similar modifications, but their origin is rather less extensive; they terminate in eight or nine rays, shorter, and with intervening angles more equal than in existing *Chelones*. The xiphisternal piece, *xs*, receives in a notch the outermost ray or spine of the inner radiated process of the hyposternal, as in the *Chelones*, and is not joined by a transverse suture, as in the *Emydes*, whether young or old.

Subjoined are dimensions of the plastron of Mr. Bowerbank's fossil:

	Inches.	Lines.
Shortest longitudinal diameter of hyosternal and hyposternal pieces	2	5
Transverse diameter of ditto	1	7
Total length of plastron	6	0

The bones of the scapular arch, especially the coracoid, Cuvier has shown to afford distinctive characters of the natural families of the *Chelonia*; but the Eocene Chelonites described by Cuvier, did not yield him this opportunity of thus testing their affinities. In the *Chelone breviceps* here described, the left coracoid (52, fig. 2) is preserved in nearly its natural position; it is long, slender, symmetrical; cylindrical near its humeral extremity; flattened, and gradually expanded from its humeral third, to its sternal end, which is relatively somewhat broader than in the *Chelone mydas* and *Chelone caouanna*.

	Inch.	Lines.
Its length is	1	6
Breadth of sternal end . . .	0	7

The characters thus afforded by the cranium, carapace, plastron, and by one of the bones of the anterior extremity, prove the present Sheppey fossil to belong to a true sea turtle; and at the same time most clearly establish its distinction from the known existing species of *Chelone*.

On account of the shortness of the skull, especially of the facial part and of that which intervenes between the orbit and ear, compared with the breadth of the skull across the mastoids, I have proposed to name this extinct species, *Chelone breviceps*.[*]

By the characteristic shape of the median extremities of the costal plates of the carapace, I have been able to determine some fragmentary Chelonites which have afforded better ideas of the size of the species represented by Mr. Bowerbank's more complete but immature specimen of *Chelone breviceps*.

A portion of the carapace of the *Chelone breviceps*, including the fourth, fifth, sixth, and part of the third and seventh neural plates, with a considerable proportion of the third, fourth, fifth, and sixth costal plates, is preserved in the museum of Mr. Robertson, of Chatham. The characters of the rugous surface of these bones, and of the equal-sided angles by which the costal plates articulate with the neural plates, do both, and especially the latter, point out the species to which the present fragment belongs. It has formed part of an individual double the size of the specimen above described, and figured from Mr. Bowerbank's collection, and therefore it had a carapace sixteen inches in length.

Although the costal plates have been continued further along the ribs than in the younger example, the more complete state of the sixth rib, in Mr. Robertson's specimen, shows that they retained their longitudinally-striated, tooth-like extremities, which, in the sixth rib, is two thirds of an inch in length; the length of the expanded part being four inches, and its breadth one inch nine lines. The internally prominent part of the rib is much less developed than in *Chelone planimentum*, and *Chelone crassicostata*, afterwards to be described. The right hyosternals and hyposternals are present, and they likewise preserve the character of the *Chel. breviceps* in their rugous surface and minor breadth, as compared with those parts in the *Chelone longiceps*, the extinct species next to be described.

Besides the specimens above described, on which the present extinct species of turtle

---

[*] Proceedings of Geological Society, December 1, 1841, p. 570. Report on British Fossil Reptiles, Trans. Brit. Association, 1841, p. 178.

has been established, remains of the *Chelone breviceps* are preserved in the Hunterian Museum, and in that of my esteemed friend and coadjutor, Professor Bell, S.R.S.

I know no other locality of the species than that of Sheppey, in Kent.

CHELONE LONGICEPS.   *Owen.* Tab. III, IV, and V.

Proceedings of Geological Society of London, December 1, 1841, p. 572.   Report on British Fossil Reptilia, Trans. British Association, 1841, p. 177.

The second species of *Chelone*, from the Eocene clay at Sheppey, which I originally recognised and defined by the fossil skull, Tab. III, differs more from those of existing *Chelones* by the regular tapering of that part into a prolonged pointed muzzle, than does the *Chelone breviceps* by its short and anteriorly-truncated cranium.

The surface of the cranial bones is smoother than in the *Chel. breviceps;* whilst their proportions and relations prove the marine character of the present fossil as strongly as in that species.

The orbits (Tab. III, figs. 1 and 2, *o,*) are large; the temporal fossæ (ib. fig. 3, *t,*) are covered principally by the posterior frontals (fig. 2, 12); and the osseous shield completed by the parietals (7), and mastoids (8), overhangs the tympanic (28), ex-occipital (2), and paroccipital (4) bones.   The compressed spine (3) of the occiput is the only part that projects further backwards.

The palatal and nasal regions of the skull afford further evidence of the affinities of the present Sheppey Chelonite to the true turtles.   The bony palate (fig. 3) presents, in an exaggerated degree, the great extent from the intermaxillary bones to the posterior nasal aperture which characterises the genus *Chelone;* and it is not perforated, as in the soft turtles (*Trionyx*), by an anterior palatal foramen.

The extent of the bony palate is relatively greater than in the *Chelone mydas,* and the trenchant alveolar ridge is less deep; the groove for the reception of that of the lower jaw is shallower than in the *Chelone mydas,* or the extinct *Chel. breviceps,* arising from the absence of the internal alveolar ridge, in which respect the *Chel. longiceps* resembles the *Chel. caretta.*

The *Chelone longiceps* is distinguished from all known existing *Chelones* by the proximity of the palatal vomer (13, fig. 3), to the basisphenoid (5), and by the depth of the groove of the pterygoid bones (24), and in both these characters in a still greater degree from the Trionyxes; to which, however, it approaches in the elongated and pointed form of the muzzle, and the trenchant character of the alveolar margin of the jaws.

The following are dimensions of the skull described :

	Inches.	Lines.
Length of the skull	4	0
Breadth of ditto across the zygomata	2	6
Antero-posterior diameter of orbit	1	2

In a second example of the skull of *Chelone longiceps*, two of the middle neural plates, and the corresponding costal plates of the right side, portions of vertebræ, with the right xiphisternal piece, humerus and femur, are cemented together, and to the cranium by the petrified clay. (T. IV, fig. 1.)

The neural plates (*s2, s3*) are flat and smooth; the entire one measures one inch two lines in length, and nine lines across its broad anterior part :—this receives the convex posterior extremity of the preceding plate in a corresponding notch A small proportion, about one sixth, of the anterior part of the external margin, joins the second costal plate (*pl2*) ; the remaining five sixths of the outer margin forms the suture for the vertebral end of the third costal plate (*pl3*).

In this respect, the *Chel. longiceps* resembles the existing *Chelones* ; and differs, as well as in the smooth and flattened surface of the vertebral plates, from the *Chelone breviceps*. The length of the third costal plate, in the fragmentary example here described, is three inches ; the impression of the commencement of the narrow portion, formed by the extremity of the coalesced rib, is preserved.

The marginal indentations of the vertebral scutes are not half a line in breadth.

The transverse impression between the first and second vertebral scute crosses the first neural plate, nine lines from its posterior extremity ; the second neural plate is free, as in other *Chelones*, from any impression, being wholly covered by the second vertebral scute.

The expanded ribs are convex at the under part, slightly concave at the upper part in the direction of the axis of the shell ; they slope very gently from the plane of the neural plates, about half an inch, for example, in an extent of three inches ; thus indicating a very depressed form of carapace.

The xiphisternal bone (*xs*), like that of *Chel. breviceps*, is relatively broader than in the existing turtles, and both the internal and external margins of its posterior half are slightly toothed. A part of the notch by which it was attached to the hyposternal remains upon the broken anterior extremity of the bone. It measures one inch two lines across its broadest part ; its length seems to have been three inches and a half.

The humerus presents the usual characters of that of the *Chelones ;* its length is two inches three lines ; its breadth across the large tuberosities ten lines. The radius and ulna extend in this Chelonite from beneath the carapace into the right orbit ; the radius is one inch and a half in length ; the ulna one inch, three lines in length ; portions of vertebræ adhere also to the mass, the state of which indicates that the animal had been buried in the clay before the parts of the skeleton had been wholly disarticulated by putrefaction.

A mass of Sheppey clay-stone supporting the ninth and tenth neural plates, and the expanded portions of the sixth, seventh, and eighth costal plates of the right side, exhibits the characters of the marine turtles in the great relative expansion of the tenth neural plate; and the tooth-like continuation of the rib from the posterior angle of the eighth costal plate (*pl8*, T. IV, fig. 2). These portions of the carapace, from their smooth surface, the impressions of the horny scutes, the form of the vertebral ends, and the concavity of the upper surface of the costal plates, evidently belong to the same species as the fossil last described.

A similar mass of Sheppey clay-stone, in Mr. Lowe's collection, supports a larger proportion of the hinder part of the carapace, including the sixth, eventh, eighth, ninth, and tenth neural plates, part of the fifth neural plate, more or less of the last four pairs of costal plates, with the impressions of the third and fourth ribs of the right side; the impression of apparently the whole of the free, slender, termination of the third rib is preserved, and also that of the fifth rib, confirming the generic characters indicated by the skull. The smooth outer surface of the bones of the carapace, the forms of the neural plates, and the concomitant modification of the commencement of the costal plates articulated therewith, concur to establish the specific distinction from the *Chelone breviceps*, and indicate the specimen to belong to the present species, *Chelone longiceps*. The seventh, eighth, and ninth neural plates progressively decrease in size; and the ninth presents a simple, quadrangular, oblong form; the tenth neural plate suddenly expands, and has apparently a triangular form, but its posterior border is incomplete.

The indications of the comparative flatness of the carapace of the *Chelone longiceps*, (in this respect, as in the elongated and pointed form of the skull, approaching the genus *Trionyx*,) which were derived from an examination of the foregoing fragments, and particularly of the portion preserved with the cranium on which the species is founded, are fully confirmed by the almost entire carapace which, subsequently to the publication of my 'Report on British Fossil Reptiles,' where the present species is first noticed, I have had the opportunity of examining in the collection of Mr. Bowerbank.

This carapace, as compared with that of the *Chelone breviceps* in the same collection, presents the following differences:—it is much broader and flatter. The neural plates are relatively broader; the lateral angle from which the intercostal suture is continued, is much nearer the anterior margin of the plate—the *Chelone longiceps*, in this respect, resembling the existing species of turtle (see fig. 1, p. 3). The costal plates are relatively longer; they are slightly concave transversely to their axis on their upper surface, while in *Chel. breviceps* they are flat. The external surface of the whole carapace is smoother; and although it is as depressed as in most turtles, it is more regularly convex; not sloping away by two nearly plane surfaces from the median longitudinal ridge of the carapace.

The following minor differences may be noticed in the two Sheppey Chelonites: the nuchal plate of the *Chel. longiceps* (Tab. V, fig. 1, *ch*) is more convex at its middle part, and sends backwards a short emarginate process to join the first neural plate (*s1*); in which it resembles the *Chel. mydas.* Both posterior angles of the first neural plates are produced, and truncate to articulate with the second pair of costal plates; and the second neural plate is quadrangular. In a portion of another carapace of the *Chelone longiceps* the second neural plate (*s2*) is pentangular, the right anterior corner being produced, and truncate to join with the first costal plate of the right side; the left posterior corner of the first neural plate (*s1*) being produced, and truncate, to articulate with the second costal plate of the left side. This structure I believe, however, to be an individual variety. But the characters of the species are exemplified in more constant modifications of the carapace. The succeeding neural plates to the seventh inclusive (*s3—s7*) are hexagonal, with the anterior lateral border much shorter than the posterior lateral border, as in *Chelone mydas*, and not of equal extent, as in *Chelone breviceps;* they become more equal in the seventh and eighth neural plates, which also decrease in size; the ninth plate (*s9*) is very small, quadrangular, and oblong, as in Mr. Lowe's fragment. Only a small portion of the tenth neural plate is preserved in Mr. Bowerbank's beautiful specimen.

The impressions of the horny scutes are deeper, and the lines which bound the sides of the vertebral scutes (*v1—v4*) meet at a much more open angle than in the *Chel. breviceps*, in which the vertebral scutes have the more regular hexagonal form of those of the *Chel. mydas.* Their relations to the neural plates are nearly the same as in *Chel. breviceps.*

The plastron (Tab. V, fig. 2) is more remarkable than that of the *Chel. breviceps* for the extent of its ossification; the central cartilaginous space being reduced to an elliptical or subquadrangular fissure. The four large middle pieces *hyosternals* (*hs*) and *hyposternals* (*ps*), have their transverse extent relatively much greater as compared with their antero-posterior extent, than in the *Chel. breviceps;* and this might be expected, in conformity with the broad character of the bony cuirass indicated by the carapace. The median margins of the *hyosternals* (*hs*) are developed in short toothed processes, along their anterior three fourths; the median margins of the *hyposternals* (*ps*) have the same structure along nearly their whole extent; the intermediate space between the smooth or edentate margins of the opposite bone is ten lines; the expanded end of the long coracoid (52) is seen projecting into this space.

The xiphisternals (*xs*) are relatively broader than in *Chel. breviceps*, or in any of the existing turtles; and are united together, or touch each other, by the toothed processes developed from the whole of their median margins. The entosternal piece is broad, flat on its under surface, and is likewise dentated at its sides.

The outer surface of each half of the plastron inclines, as in the *Chelone mydas*, towards a submedian longitudinal ridge.

The breadth of the plastron, in the specimen figured (fig. 2), along the median suture, uniting the hyosternals and hyposternals, is six inches: the narrowest antero-posterior diameter of the conjoined hyosternals and hyposternals is two inches nine lines.

The breadth of the plastron, at the junction of the xiphisternals with the hyposternals, is two inches six lines.

The posterior part of the cranium is preserved in Mr. Bowerbank's specimen (fig. 1), withdrawn beneath the anterior part of the carapace ; the fracture shows the osseous shield covering the temporal fossæ; and the pterygoids remain, exhibiting the deep groove that runs along their under part.

It is most satisfactory to have found that the two distinct species of the genus *Chelone*, determined, in the first instance, by the skulls only, should thus have been confirmed by the subsequent comparison of their bony cuirasses ; and that the specific differences, manifested by the cuirasses, should be proved by good evidence to be characteristic of the two species founded on the skulls.

Thus the portion of the skull preserved with the carapace first described (Tab. II, figs. 2 and 3), served to identify that fossil with the more perfect skull of the *Chelone breviceps* (Tab. I), by which the species was first indicated. And, again, the portion of the carapace adhering to the perfect skull of the *Chelone longiceps* (Tab. IV, fig. 1) equally served to connect with it the nearly complete osseous buckler (Tab. V, fig. 1), which, otherwise, from the very small fragment of the skull remaining attached to it, could only have been assigned conjecturally to the *Chel. longiceps;* an approximation which would have been the more hazardous, since the *Chelone breviceps* and *Chelone longiceps* are not the only turtles which swam those ancient seas that received the enormous argillaceous deposits of which the Isle of Sheppey forms a part.

CHELONE LATISCUTATA. *Owen.* Tab. VI.
Proceedings of the Geological Society of London, December 1, 1841, p. 574. Report on British Fossil Reptiles, Trans. British Association, 1841, p. 179.

A considerable portion, measuring three inches in length, of the bony cuirass of a young turtle from Sheppey, including the first to the sixth neural plates (T. VI, fig. 1, $s1$—$s6$), with the corresponding pairs of costal plates ($pl1$—$pl6$), and the hyosternal (fig. 2, hs) and hyposternal (ps) elements of the plastron, most resembles that of the *Chelone longiceps* in the form of the carapace, and especially in the great transverse extent of the above-named parts of the plastron: it differs, however, from the *Chel. longiceps*, and the other known fossil Chelonites, in the greater relative breadth of the vertebral scutes ($v2, v3$), which are nearly twice as broad as they are long.

The central vacuity of the plastron is subcircular; and, as might be expected, from the apparent nonage of the specimen, is wider than in the *Chel. longiceps;* but the toothed processes given off from the inner margin of both hyosternals and hyposternals are small, sub-equal, regular in their direction, and thus resemble those of the *Chel. longiceps;* the slender point of the episternal (s) is preserved in the interspace between the hyosternals. Both hyosternals (*hs*) and hyposternals (*ps*) are slightly bent upon a median longitudinal prominence of their under surfaces.

The length of the third costal plate (*pl3*) is one inch seven lines; its antero-posterior diameter or breadth, six lines: in the form of the vertebral extremities of the costal plates, and of the neural plates to which they are articulated, the present fossil resembles the *Chel. longiceps*; but the fifth neural plate is more convex, and is crossed by the impression dividing the third vertebral scute (*v3*) from the fourth, which impression crosses the suture between the fifth and sixth neural plates in both *Chelone longiceps* and *Chelone breviceps*. Whether, in the progressive change of form, which the vertebral scutes may have undergone in the growth of this young turtle, as during the growth of the young loggerhead turtle (*Chelone caouanna*), by an increase of length, without corresponding increase of breadth, the impression between the third and forth vertebral scute, might also retrograde to the interval between the fifth and sixth neural plates, I am uncertain, having only had the opportunity of comparing the scutes of the young and old loggerhead turtles, not the skeletons. The change in the lateral angles of the vertebral scutes, resulting from the elongation of the scutes themselves, in the loggerhead, would be similar to that in the *Chelone longiceps*, as compared with the *Chel. latiscutata*, on the hypothesis that the latter is the young of the former; but in my present uncertainty I prefer to indicate the specimen in question, by the definite name proposed in my original Memoir; its description as a distinct species being more likely to attract the attention of Collectors to similar specimens, and to enable them to identify such. Figure 3, T. VI, gives the degree of convexity of the carapace, and the double curve of the plastron produced by the prominence of the principal hæmapophyses *hs* and *ps*. The left scapular arch (51) is exposed in this view.

CHELONE CONVEXA. *Owen.* Tab. VII.

Proceedings of the Geological Society of London, December 1, 1841, p. 575. Report on British Fossil Reptiles, Trans. British Association, 1841, p. 178.

The fourth species of *Chelone*, indicated by a nearly complete cuirass, from Sheppey, holds a somewhat intermediate position between the *Chelone breviceps* and the *Chelone longiceps*; the carapace being narrower, and more convex than that of *Chel. longiceps*; broader and with a more regular transverse curvature than in the *Chelone breviceps*.

Although the specimen is equal in size to either of the two with which it is here compared, the costal plates hold an intermediate length, which shows that this character is not due to a difference depending upon age.

The fossil in question includes the first to the eighth neural plate inclusive; the first plate (*s1*) expands behind, and both posterior angles are truncated to articulate with the second costal plates (*pl2*). The second neural plate (*s2*) is quadrate, half as long again as broad, and the second pair of costal plates articulate with this, as well as with the first and third plates, as in the *Chel. longiceps* (Tab. V, fig. 1). The tooth-like extremity of the connate rib is preserved on the right side. The fourth costal plate (*pl4*) is two inches four lines in length, nine lines in breadth; the angle at which the expanded part contracts to the extremity of the connate rib is well shown on the right side. The third to the eighth neural plates expand anteriorly, and have the anterior angles cut off to articulate with the costal plates in advance; they diminish in size very gradually, and the antero-lateral borders, formed by the above-named truncated angles, do not increase in length as in the corresponding plates in the *Chelone longiceps*.

The vertebral scutes (*v2, v3, v4*) resemble more in form those of the *Chel. longiceps* than of *Chel. breviceps*; but, notwithstanding that the whole carapace is narrower than in *Chel. longiceps*, the vertebral scutes are broader; and the lines which converge to the lateral angle have a more marked sigmoid curvature.

	Chel. convexa.		*Chel. longiceps.*	
	Inches.	Lines	Inches.	Lines.
The length of the second vertebral scute is	1	8	1	8
Breadth	2	6	2	2

The two succeeding scutes (*v3* and *v4*) more rapidly diminish in size than in either the *Chel. breviceps* or *longiceps*, and the transverse impression between the third and fourth vertebral scute crosses the lower third of the fifth neural plate, as in *Chelone latiscutata*. All the scutes have left deeper and rather wider impressions than in the preceding species.

The second to the fifth costal plates inclusive, are more equal in length than in the existing *Chelone mydas* or *Chel. caouanna*, and in this character the present species more resembles the *Chel. imbricata*.

The distinction of the present from the previously described fossils, already manifested in the structure of the carapace and the form of the vertebral scutes, is more strikingly established in that of the plastron (Tab. VII, fig. 2), which, in its defective ossification, resembles the same part in the existing species of *Chelone*.

All the bones, but especially the xiphisternals (*xs*), are more convex on their outer surface than in other turtles, recent or fossil. The central vacuity is greater than in any of the above-described fossil species. The internal rays of the hyosternals come off from the anterior half of their inner border, and are divided into two groups: the lower consisting of two short and strong teeth, projecting inwards towards the extremity of the entosternal (*s*); while the rest extend forwards along the inner side of episternals (*es*). The same character may be observed in the corresponding processes of the *hyposternals* (*ps*), which are limited to the posterior half of their inner border. The external radiated process of the hyosternals (*hs*) arises from a larger proportion of the outer margin, than in the *Chel. mydas*; but from a somewhat less proportion than in *Chel. breviceps*.

The external process of the hyposternal (*ps*) is relatively much narrower than in the *Chel. breviceps* (T. II, fig. 2), and, *à fortiori*, than in *Chel. longiceps* (Tab. V, fig. 2). The straight transverse suture by which the hyosternals and hyposternals of the same side are joined together, is much shorter than in the other fossil *Chelones*; and is similar in extent to that in *Chel. mydas*; but the following differences present themselves in the plastron of the *Chelone convexa*, as compared with that of the *Chelone mydas*.

The median margin of the hyosternals forms a gentle curve, not an angle: that of the hyposternals is likewise curved, but with a slight notch. The longitudinal ridge on the external surface is nearer the median margin of the *hyosternals* and *hyposternals* and is less marked than in the *Chelone longiceps*; especially in the hyposternals, which are characterised by a smooth concavity in the middle of their outer surface.

The suture between the *hyosternals* and *hyposternals* is nearer to the external, transverse, radiated process of the hyposternals. The median vacuity of the sternal apparatus is elliptical in the *Chel. convexa*, but square in the *Chel. mydas*.

The characteristic lanceolate form of the episternal bone (*s*) in the genus *Chelone*, is well seen in the present fossil. The entosternal element of the plastron is subcircular, or lozenge-shaped; and generally broader than it is long in the Emydians.

The true marine character of the present Sheppey Chelonite, so well given in the carapace and plastron, is likewise satisfactorily shown in the small relative size of the entire femur (65) which is preserved on the left side, attached by the matrix to the left xiphisternal. It presents the usual form, and slight sigmoid flexure, characteristic of the *Chelones*; it measures one inch in length.

In an *Emys* of the same size, the femur, besides its greater bend, is one inch and a half in length.

A Chelonian cranium from Sheppey, two inches five lines in length, in the museum of Professor Bell (T. VI, fig. 4), and a second of the same species from the same locality, two inches nine lines in length, in the museum of Fred. Dixon, Esq., F.G.S., of Worthing, belong to the same species, and differ from the cranium of the *Chelone breviceps*, in the more pointed form of the muzzle, and the less rugose character of the outer surface of the bones; they equally differ from the *Chelone longiceps* in the less produced, and less acute muzzle, and the more rugose surface of the bones. The parietals (7) are bounded anteriorly by a semicircular line, not by a semioval one, as in *Chel. longiceps*, or by an angular one, as in *Chel. breviceps*. The frontals (11) enter into the formation of the orbits, as in both the foregoing species. The orbits are subcircular, as in *Chel. longiceps*, not subrhomboidal with the angle rounded off, as in *Chel. breviceps*. The postfrontals (12) are large, and form a slight projection at the back part of the supraorbital ridge. The tympanic cavity is larger in proportion than in the *Chelone longiceps*. The palate is traversed by a deep median, longitudinal groove, between which and the shallower grooves on the inner sides of the alveolar borders, are two well-marked, diverging, longitudinal prominences. The bony palate is longer than in *Chelone breviceps*, shorter than in *Chel. longiceps*. The symphysis of the lower jaw (T. VII, fig. 3) is longer or deeper than in the *Chelone breviceps*, but is convex below from side to side, and not flattened as in the *Chelone planimentum*.

All the specimens of *Chelone convexa*, which I have been able to determine, are from the London clay of Sheppey.

CHELONE SUBCRISTATA.   *Owen.*   Tab. VIII.

Proceedings of the Geological Society of London, December 1, 1841, p. 576.   Report on British Fossil Reptiles, Trans. British Association, 1841, p. 179.

The fifth species of *Chelone* from Sheppey, distinguishable by the characters of its carapace, approaches more nearly to the *Chelone caouanna* in the form of the vertebral scutes ($v1$—$v4$), which are narrower in proportion to their length, than in any of the previously described species; but the *Chelone subcristata* is more conspicuously distinct by the form of the fifth and seventh neural plates ($s5$, $s7$), each of which supports a short, sharp, longitudinal crest; a similar crest is developed from the contiguous ends of the second and third neural plates ($s2$, $s3$); the middle and posterior part of the nuchal plate (ch) is raised into a convexity, as in the *Chel. longiceps;* but not into a crest.

The keeled structure of the above-cited neural plates is more marked than in the third and fifth neural plates of *Chelone mydas*, which are raised into a longitudinal ridge.

The neural plates in the present carapace have the ordinary, narrow, elongated form of those in the true *Chelones*.   The nuchal plate (ch) has the middle of its hinder border produced backwards, instead of being emarginate, as in the *Chel. breviceps* (T. II, fig. 1, ch).

The first neural plate in the *Chelone subcristata* (T. VIII, $s1$) resembles that in the *Chelone convexa*, but is narrower in proportion to its length; the second ($s2$) is also quadrangular, as in *Chel. convexa*, but is narrower; the third to the seventh likewise differ from those in *Chel. convexa* only by being narrower; but the eighth and ninth neural plates are relatively smaller than in any of the before-described fossils, and resemble those of existing *Chelones*.   The expanded plate is more elevated, and is bent down on each side, with the middle part forming an obtuse longitudinal ridge. A part of the contiguous portion of the first ($pl1$) and the second ($pl2$) costal plates are raised into a slight convex eminence on each side; the surface of the remaining pairs of ribs is flat in the axis of the body, but they are more convex transversely to that axis, and in the direction of their own length, than in the other Chelonites.

The whole outer surface of the bones of the carapace is as smooth as in the *Chel. longiceps* and *Chel. convexa*.

Length of carapace from the first to the eighth neural plate inclusive:

Ch. subcristata.		Ch. breviceps.		Ch. longiceps.		Ch. convexa.	
Inches	Lines.	Inches	Lines.	Inches.	Lines.	Inches.	Lines.
7	4	5	6	5	9	5	8

The length of the present fossil carapace, to the tenth neural plate, inclusive, is nine inches.

The breadth between the ends of the third costal plates, in a straight line, is six inches six lines. The succeeding costal plates more gradually decrease in breadth, than in the *Chel. longiceps* and *Chel. convexa;* and the entire carapace more resembles in form that of the *Chel. mydas,* and *Chel. caouanna.*

The epidermal scutes are defined by deep impressions, and as wide, relatively, as in the *Chel. mydas* and *Chel. convexa.* The length of the second vertebral scute is two inches one line; its breadth is two inches two lines; the length of the fourth vertebral scute is two inches three lines; and its breadth one inch eleven lines, and, at its posterior margin, only nine lines. This scute is narrower than in *Chel. caouanna,* or any of the previously described fossil species; the outer angles are less produced than in the *Chelone caouanna.*

Sufficient of the plastron is exposed in the present fossil to show by its narrow elongated xiphisternals (*xs*), and by the wide and deep notch in the outer margin of the conjoined hyosternals and hyposternals (*hs* and *ps*), that it belongs to the marine *Chelones.* The xiphisternals are articulated to the hyposternals by the usual notch or gomphosis; they are straighter and more approximated than in the *Chel. mydas* and *Chel. caouanna.* The external emargination of the plastron between the hyosternals and hyposternals, differs from that of the recent turtles in being semicircular, instead of angular; the *Chel. subcristata* approaching, in this respect, to the *Chel. breviceps.* The shortest antero-posterior diameter of the conjoined hyosternals and hyposternals is two inches seven lines. The length of the xiphisternal is two inches six lines; the breadth of both, across their middle part, is one inch three lines.

The name proposed for this species indicates its chief distinguishing character, viz., the median interrupted carina of the carapace, which may be presumed to have been more conspicuous in the horny plates of the recent animal, than in the supporting bones of the petrified carapace.

CHELONE PLANIMENTUM. *Owen.* Tab. IX and X.
Proceedings of the Geological Society of London, December, 1841, p. 576. Report on British Fossil Reptiles, Trans. British Association, 1841, p. 178.
Syn. CHELONE HARVICENSIS, *Woodward* (?).

The skull of a large *Chelone* (T. IX) from the Eocene clay near Harwich, in Professor Sedgwick's collection at Cambridge, resembles, in the pointed form of the muzzle, the *Chel. longiceps* of Sheppey; but differs in the greater convexity and breadth of the cranium (fig. 2); and the more abrupt declivity of its anterior contour (fig. 3), and from other *Chelones* by the broad expanse of the inferiorly-flattened *symphysis menti* (fig. 1).

4

The osseous roof of the temporal fossæ, and the share contributed to that roof by the postfrontals (T. IX, figs. 2 and 3, 12), distinguish the present, equally with the foregoing Chelonites, from the *Emys* (*Podocnemys*) *expansa*, and, *à fortiori*, from other genera and species of the fresh-water families (*Emydidæ* and *Trionicidæ*).

In the oblique position of the orbits (fig. 3, *o*), and the diminished breadth of the interorbital space (fig. 2), the present Chelonite, however, approaches nearer to *Trionyx* and *Emys* than do the previously-described species. But the sides of the face converge more rapidly towards the muzzle. Its most marked and characteristic difference from all existing *Chelones* is shown by the greater antero-posterior extent, breadth, and flatness of the under part of the symphysis of the lower jaw, whence the specific name here given to the species. The posterior border of the symphysis is defined by a regular semicircular curve, and the rami of the jaw have completely coalesced.

Since at present there is no means of identifying the well-marked species, of which the skull is here described, with the Chelonite figured in the frontispiece to Woodward's 'Synoptical Table of British Organic Remains,' and alluded to, without additional description or characters, as the *Chelonia Harvicensis*, in the additions to Mr. Gray's 'Synopsis Reptilium' (p. 78, 1831); and since the extensive deposit of Eocene clay along the coast of Essex, like that at the mouth of the Thames, contains the relics of more than one species of ancient British turtles,* I prefer indicating the one here established by a name having reference to its peculiarly distinguishing character, rather than to associate arbitrarily the skull, which gives the true specific distinction, with the ill-defined carapace to which the vague name of *Harvicensis* has been applied; more especially as the fossil carapace to which the present skull more probably belongs, from the circumstance under which it was discovered, also presents well-marked, and readily-recognisable specific characters.

This carapace (T. X) is also contained in the museum of Professor Sedgwick, and is understood to have formed part of the same individual turtle as the skull (T. IX) on which the species, *Chel. planimentum*, was founded.

In general form this carapace differs from that of the existing *Chelones*, in being less contracted and pointed posteriorly than in the *Chelone mydas* and *Chel caouanna*, and more contracted posteriorly than in the *Chel. imbricata*. In the proportion which the pleurapophyses (true ribs), bear to the superimposed costal plates, (*pl*4—8) it resembles *Chelone mydas*, and *Chelone caouanna*, more than it does the *Chel. imbricata*. But the pleurapophyses are more prominent and distinct from the costal plates throughout their entire length, than in the *Chel. mydas* or *Chel. caouanna*, and present an obtuse angular ridge towards the cavity of the abdomen.

The five posterior pairs of ribs of the carapace (*pl*4—*pl*8) are preserved, with part

---

* Sir C. Lyell alludes to the Chelonites of Harwich in his 'Elements of Geology:' "This formation is well seen in the neighbouring cliffs of Harwich, where the nodules contain many marine shells, and sometimes the bones of Turtles." (Vol. ii, p. 337.)

of the first three on the left side, and one of the coracoids (52) showing the rather sudden and considerable expansion of its sternal or mesial half.

The interval between the free extremities of most of the ribs, is about equal to twice and a half the breadth of each extremity; but the interval between the seventh (*pl7*) and eighth (*pl8*) rib, measured, like the others, at the terminal border of the costal plates, is equal to thrice the breadth of the free part of the seventh rib.

In this respect the *Chelone planimentum* resembles the *Chel. mydas* more than it does the *Chelone caouanna*, in which the interval between the free extremities of the seventh and eighth ribs is less than that between the sixth and seventh. The length of the costal plate of the fourth rib is twice that of the eighth rib, as in the *Chelone caouanna;* in *Chel. mydas* it is more than twice as long; in *Chel. imbricata* it is only one third longer. The marginal pieces in the *Chelone planimentum* seem to have been narrow or slender in proportion to their length.

The following admeasurements show that, in the large proportionate size of the head, the *Chelone planimentum* corresponds with the existing turtles :

	Inches.	Lines.
Length of the cranium	5	6
Depth of ditto	4	0
Breadth of ditto	5	0
Length of the carapace	15	6
Greatest breadth of ditto	13	0

Tab. IX and X satisfactorily illustrate the characteristic forms and proportions of the unique specimen in the Cambridge Museum; the carapace is figured of half the natural size.

CHELONE CRASSICOSTATA. *Owen.* Tab. XI and XII.
TESTUDO PLANA. *König.* ' Icones Sectiles,' Pl. XVI, fig. 192 ?

That the extinct species of Eocene turtles attained larger dimensions than those given above, is proved by a fossil skull from the Harwich clay, in the collection of Professor Bell, which gives the following dimensions :

	Inches.	Lines.
Total length of the cranium	8	0
Its greatest breadth	6	0
The antero-posterior extent of the *symphysis menti* . . .	3	0
The vertical diameter of the orbit	1	9
do. do. of the nostri . , . . .	0	9

This skull differs from that of the *Chelone planimentum* in the minor depth of the maxillary bone below the orbit (compare T. IX, fig. 3, with T. XI, fig. 2, 21), in the more acute and attenuated muzzle; but especially in the minor breadth and the different configuration of the posterior margin of the *symphysis* of the lower jaw (compare T. IX, fig. 1, with T. XI, fig. 3). With regard to the comparative anatomy of the bones of the skull, and the pattern of the scutation of the upper surface of the cranium, I regret that the state of the specimen in Professor Bell's collection does not permit the deduction of other distinctive characters which such parts of the cranial organization so satisfactorily afford. A great proportion of the osseous parietes is wanting; but the cast in the hard matrix of the wide lateral cavities (12, 12), which were over-arched by the expanded postfrontal and parietal bones, indicates the prominence of the postfrontals at the upper and outer angle of the orbits. The orbits (*or*) appear to have been more ovate and less circular than in the *Chelone planimentum*; and the sides of the orbital part of the skull do not converge so rapidly towards the muzzle, but meet at a more acute angle.

That a second species of turtle, distinct from the *Chelone planimentum*, has left its remains in the Harwich clay, is very decisively demonstrated by the almost complete carapace in the British Museum, the inner surface of which is represented, on the scale of six inches to a foot, in T. XII. This carapace, both by its general contour, by the relative length of the costal plates to one another, and by their relative breadth to the adherent pleurapophyses beneath, more resembles the carapace of the *Chelone imbricata* than that of the other known existing species of turtle; and, as the peculiar characters of the *Chelone imbricata* are exaggerated, it differs in a proportional degree from the *Chelone planimentum*. These characters are seen in the great breadth of the prominent inferior part of the ribs, and of the free extremity of the rib (*pl*1—*pl*8), as compared with the total breadth of the costal plate. The intervals between the free extremities, where the expanded plate terminates, are not equal to the breadth of the proper ribs; in the *Chelone imbricata* they very slightly exceed the breadth of the free ends of the ribs. This character in the fossil, by which it is so markedly distinguished from the *Chelone planimentum*, and most other species, has suggested the name *Chelone crassicostata*, or thick-ribbed turtle, which is proposed for the present species. The last pair of ribs of the carapace (T. XII, *pl*8) are remarkably short and thick, and are curved backwards on each side the broad terminal neural plates which they almost touch. In this character the *Chel. crassicostata* resembles the *Chel. imbricata*, and differs from the *Chel. caouanna* (fig. 2, p. 3), and from *Chel. mydas*. The subequality of length of the costal plates is another character by which the *Chel. crassicostata* resembles the *Chel. imbricata*, and differs from the *Chel. mydas*, the *Chel. caouanna*, as well as from the *Chel. planimentum*.

In T. XII, as in the other figures, *ch* is the nuchal plate, *pl*1 the first rib of the carapace (the second free pleurapophysis or vertebral rib), *pl*2 to *pl*8 the remaining ribs of the carapace and costal plates; *s*9, *s*10, and *py* are the terminal neural plates and pygal plate, which, like the nuchal plate, are developed in the substance of the integument, without becoming attached to the subjacent spinous processes of the vertebræ. The debris of the neural arches of the intermediate eight vertebræ of the carapace are preserved in the interspaces of the beginnings of the ribs and costal plates in this beautiful Chelonite. It forms part of the Fossil Collection in the British Museum.

A carapace of a smaller individual of *Chelone crassicostata*, from the Harwich coast, with the character of the broad and inwardly-prominent ribs strongly marked, is likewise preserved in the choice collection of my esteemed friend Professor Bell. One of the hyosternal bones, inclosed in the same nodule of clay, testifies to the partial ossification of the plastron in this species by its coarsely-dentated border; and, at the same time, shows a specific peculiarity by the convexity of that surface which was turned towards the cavity of the thoracic-abdominal case. On the moiety of the nodule containing the carapace and exposing its under surface, the slender rudimental rib of the proper first dorsal vertebræ is preserved, in connexion with the first expanded rib of the carapace.

Besides the specimen of *Chelone crassicostata* from Harwich, figured in T. XII, there is a mutilated carapace of a young *Chelone*, from the same locality, in the British Museum. This specimen exhibits the inner side of the carapace, with the heads, and part of the expanded bodies, of four pairs of ribs, which indicate its specific agreement with the foregoing specimen, and demonstrate unequivocally its title to rank with the marine turtles. It is figured in Mr. Kœnig's '*Icones Sectiles*' (pl. xvi, fig. 192), under the name of *Testudo plana*.

A rare Chelonite from the hard Eocene clay apparently of Harwich, in the collection of my friend Frederick Dixon, Esq., F.G.S., of Worthing, shows the impressions from the under surface of the carapace, and also an instructive part of the under surface of the plastron itself. (T. XIII.) The proportions and degree of convexity of the under surface of the costal plates of the carapace (*pl, pl*) correspond with those parts in the *Chelone crassicostata*.

The remains of the plastron include a great portion of the left hyosternal (*hs*), left hyposternal (*ps*), and left xiphisternal (*xs*); the latter is articulated to the hyposternal by a notch, receiving a toothed process, and, reciprocally, near the upper part of a long oblique harmonia, between the outer border of the hinder angle of the hyposternal and the inner border of the upper half of the xiphisternal. The hyosternal is concave lengthwise, and is convex across on its under surface; the transverse linear impression, dividing the pectoral and abdominal scutes, crosses near its posterior border. The degree of concavity of the outer surface of this bone corresponds with the convexity of the upper and inner surface of the same bone in the specimen of the *Chelone crassicostata* from Harwich, in the Museum of Professor Bell; and it concurs with the characters of the costal plates in proving the present Chelonite to be of the same species. Impressions of the toothed mesial margin of the right hyosternal remain, and part of the toothed margin of the left hyposternal.

The right coracoid (52) is exposed by the removal of the right hyosternal; it differs in form from that preserved in the large specimen of *Chelone planimentum*, in Professor Sedgwick's Museum, in expanding less suddenly at its sternal end, as compared with the coracoid of the *Chelone mydas*, or with that of the *Chelone caouanna*, which is somewhat broader than in the *Chel. mydas;* the coracoid of the *Chel. crassicostata* agrees with that of the *Chel. planimentum* in the greater degree of its expansion. At the anterior fractured surface of Mr. Dixon's Chelonite, the long and slender columnar or rib-like scapula, is shown, extending from the under part of the head of the second costal rib downwards and outwards, for an extent of two inches, and then sending its acromial or clavicular prolongation at the usual open angle downwards and inwards to rest upon the episternal. The proportions of these parts of the scapular arch are quite those which characterise the genus *Chelone,* but they do not supply such marks of specific distinction as the coracoid element does.

CHELONE DECLIVIS. *Owen.* Tab. XIV.

The extinct turtle represented by this specimen, and indicated by the above term, bears the same relation to the *Chelone convexa,* which the *Chelone longiceps** does to the *Chelone latiscutata* ;† that is, it has the same general characters of the petrified parts of the carapace, but differs in the narrower proportions of the vertebral scutes ($v1$—$v4$), and the more open angle at which their two lateral borders meet; the vertebral angles of the costal scutes being correspondingly less acute.

The specimen is from the Eocene deposits of Bognor, Sussex, and is preserved in the collection of Frederick Dixon, Esq. It consists of the seven anterior neural plates, and the corresponding seven pairs of costal plates (T. XIV), those of the right side having been broken away from their attachments to the neural plates, and bent upon the rest of the carapace at an acute angle with some slight separation of the sutures of the costal plates (fig. 2).

The neural plates correspond in general form with those of the *Chelone convexa,* the hind ones being rather broader; the first ($s1$) is crossed at its middle part by the impression dividing the first ($v1$) from the second ($v2$) vertebral scute; the second neural plate ($s2$) is an oblong four-sided one, with both ends of equal breadth. The third neural plate, $s3$, resumes the hexagonal figure with the broadest end, and two shortest sides at the fore part; and is crossed in its lower half by the impression dividing the second, $v2$, from the third vertebral scute, $v3$. The fifth neural plate ($s5$) is crossed by the next transverse impression nearer its lower border. The sixth and seventh neural plates retain the same form and proportions as in the *Chelone convexa,* except a somewhat

* Proceedings of the Geological Society of London, December 1, 1841, p. 572.
† Ibid., p. 574.

greater breadth, and have not their antero-lateral borders increased in length, as in the *Chelone longiceps.*

The declination of the ribs from the neural plates, gives a greater degree of steepness to the sides of the carapace than in the *Chelone convexa,* and the impressions of the scutes have equal depth and breadth. The chief difference indicative of specific distinctions, lies in the form of those impressions; and the question is, whether, in the progress of growth which makes the longitudinal extent of two of the vertebral scutes in one specimen nearly equal to three, in another, so great a change could be effected in their shape as is shown in the specimen of *Chelone convexa;* in which it will be seen that the second vertebral scute (T. VII, *v*2), though more than one third shorter than in *Chel. declivis* (T. XIV, *v*2), is of the same breadth as that in the larger specimen, and that the rest differ in the same remarkable degree.

CHELONE TRIGONICEPS. *Owen.*

More than one of the old tertiary turtles (*Chelone*) are remarkable for the longitudinal extent or depth of the symphysis of the lower jaw.

The turtles from the Eocene clay at Harwich have this character so strongly developed and the under surface of the symphysis so flattened, especially in one of the species, as to have suggested the " nomen triviale" *planimentum* for it. The *Chelone longiceps,* if we may judge by the length of the upper jaw and bony palate, must have had a corresponding extent of the symphysis of the under jaw; and we may infer the same peculiarity from the straight alveolar borders of the maxillaries and their acute convergence towards the premaxillary bones in an allied species, *Chelone trigoniceps,* which I have described and figured in the Appendix to Mr. Dixon's work on the 'Fossils of Sussex,' from a specimen which is in the collection of G. A. Coombe, Esq., and which was obtained from the Eocene clay at Bracklesham.

Amongst the Chelonites which Mr. Dixon has obtained from the same formation and locality, are portions of the fore part of the lower jaw of four individuals of the genus *Chelone,* all exhibiting the characters of the pointed form and great depth of the symphysis.

One of these specimens agrees so closely in size and shape with the fore part of the upper jaw of the *Chelone trigoniceps*—fits, in fact, so exactly within the alveolar border, and so closely resembles that specimen in texture and colour, that, coming from the same formation and locality, and being obtained by the same collectors, I strongly suspect it to belong to the same species of *Chelone,* if not to the same individual.

The known recent *Chelones* differ among themselves in the shape and extent of the bony symphysis of the lower jaw. Both the *Chelone imbricata,* and *Chelone caouanna* have this part deeper and more pointed than the *Chel. mydas,* but neither species has the symphysis so depressed or so slightly convex below as it is in the Bracklesham *Chelones*.

These also differ amongst themselves in this respect. The symphysis which I have referred to the *Chelone trigoniceps*, is the broadest and flattest; at its back part it shows a deep and broad genio-hyoid groove; this is reduced to a transversely oblong foramen in *Chelone mydas*.

The second species from Bracklesham, is indicated by the maxillary symphysis, the sides of which meet at a more acute angle, and it is narrower in proportion to its length, is more convex below, and more concave above, with the alveolar borders a little more raised, and the middle line less raised than in *Chelone trigoniceps*. In this respect it is intermediate between the *Chelone imbricata*, where the upper surface of the symphysis is more concave, and the *Chelone caouanna*, where it is flatter than in the *Chelone trigoniceps*. The fossil symphysis under notice, has also a smooth, transverse, genio-hyoid groove at its back part. It accords so closely in form with the end of the upper jaw of the *Chelone longiceps*, from Sheppey, that I refer it provisionally to that species.

Two other specimens of the symphysis of the lower jaw, of rather larger size, appear to belong to the same species as that referred to the *Chel. longiceps*, by the characters of the concavity of the upper surface, the convexity of the lower surface, and the degree of convergence of the sides or borders of the symphysis. The larger of the two shows the genio-hyoid groove, and the nearly vertical outer side of the jaw, opposite the back part of the symphysis, and this shows no impression of the smooth fossa receiving the insertion of the biting muscles, whereas, in the *Chelone trigoniceps*, fig. 11, that fossa extends to the same transverse line or parallel with the back part of the symphysis.

The very rare and interesting Chelonite in Mr. Coombe's museum, was the first portion of the cranium of a reptile of this order that I had seen from the Eocene deposits at Bracklesham. It includes the bones forming the roof of the mouth, with portions of the bony nostrils and orbits, and the tympanic pedicles.

The extremity of the upper jaw is broken off, but the straight converging alveolar borders clearly indicate the muzzle to have been pointed, as in the *Chelone longiceps* of Sheppey; and the muzzle being shorter, the form of the skull has more nearly approached that of a right-angled triangle. The whole cranium is broader and shorter, and the tympanic pedicles wider apart. The middle line of the palate developes a somewhat stronger ridge; the orbits were relatively larger and advanced near to the muzzle: the malar bones are more protuberant behind the orbits, and their external surface inclines inwards as it descends from behind and below the orbit, to form the lower border of the zygoma, which it does not do in the *Chelone longiceps*.

The upper surface of the fossil shows the palatines rising to form the vomer at the middle line, and the two small subcircular vacuities (occupied by membrane in the recent skull) between the palatines, prefrontals, and maxillaries; the anterior border of the temporal fossa, formed by the malar and pterygoid, is entire on one side, and shows that that vacuity was as broad as it is long. The olfactory excavations in the maxillaries are deep. The articular surface of the tympanitic pedicles closely accords with those of recent *Chelones*. The very regular triangular form of the skull indicated by this fragment, has induced me to propose the name of *Chelone trigoniceps* for the species.

CHELONE CUNEICEPS. *Owen.* Tab. XV.

One of the most complete and instructive crania of the fossil turtles of our Eocene deposits is the subject of T. XV, the opportunity of describing and figuring which has been kindly afforded me by J. Toulmin Smith, Esq., F.G.S., of whose cabinet it forms part, and by whose skilful manipulation its variously configurated exterior has been disencumbered of the hard adherent clay.

From the *Chelone breviceps* this specimen differs by its more prolonged and pointed muzzle; by the more sudden and sloping declivity of the prefrontal part of the cranium (fig. 1, 14); by the minor degree of rugosity of the surface of the bones ; and by the different disposition of the superincumbent horny scutella, which is indicated by their impressions. In the general arrangement of these impressions it accords better with the cranium of the *Chelone longiceps;* but differs in the greater breadth of the skull as compared with its length ; in the minor extent of the bony palate (fig. 3, 20, 21), the more advanced position of the posterior nostrils, and the greater length of the pterygoids (24). From the *Chelone convexa* it differs, in the greater relative breadth and flatness of the frontal bones, and of the whole interorbital platform (fig. 2, 11), in the downward slope of that part of the cranial profile, and in the more prominent convexities of the palatal processes of the maxillaries. From the *Chelone planimentum* it differs also, by the broader prefrontal part of the interorbital space, as compared with the transverse diameter of the back part of the skull; by the minor degree in which the frontal enters into the formation of the upper rim of the orbits ; by the minor depth of the suborbital part of the maxillary and malar bones, and by a very different arrangement of the supracranial horny scutella.

The basi-occipital (T. XV, figs. 3 and 4) is remarkable for the strong development of the tubercles for the insertion of the strong " recti capitis antici," and for the depth of the median groove between them ; the semicircular fossa in front of these processes is bounded by a well-developed basi-sphenoidal ridge (5), the curve of which is deeper than in *Chel. longiceps*, but shallower than in *Chel. breviceps*. In the *Chel. caouanna*, in which the basi-occipital tuberosities are better developed than in the *Chel. imbricata* or *Chel. mydas*, they are bounded anteriorly by an angular or chevron-shaped ridge of the basi-sphenoid. The exoccipitals (2) form the usual share of the trilobate occipital

5

condyle characteristic of the *Chelonia*. The paroccipitals (4) project backwards to a little beyond the posterior plane of the condyle, indicating an affinity to the *Trionycidæ*. The inferior surface of the part of the tympanic to which they unite is concave. The parietals (fig. 2, 7) form together a large semielliptic, almost flattened, platform, relatively broader than in *Chel. mydas*, not convex, as in *Chel. caouanna;* not indented by the mastoids, as in the *Chel. longiceps*, and not forming an angle between the frontals and postfrontals, as in the *Chel. breviceps*. The frontals (11) together form a pentagon, with the longest margin joining the parietals, the next in length converging to a point between the prefrontals, and the shortest borders joining the postfrontals. The postfrontals (12) and prefrontals (14) almost meet above the orbits, and exclude the frontals from entering into the formation of its superior border. The *Chel. mydas* comes nearest to the *Chel. cuneiceps* in this particular; whilst in the *Chel. imbricata* the frontals enter as largely into the formation of the upper border of the orbit as they do in the *Chel. breviceps*, *Chel. longiceps*, and *Chel. convexa*.

The precise form of the termination of the prefrontonasals, the maxillaries, and premaxillaries cannot be determined in the present specimen; fortunately, the fracture of the anterior extremity of the skull has not extended to that of the bony palate. If this be bounded by a transverse line behind, drawn across the anterior border of the temporal fossæ, the space included forms a right-angled triangle, and includes the whole of the posterior nostrils. In the *Chel. longiceps* the similarly defined space has the base shorter than the converging sides, and the posterior nasal aperture is behind the transverse line. The bony palate, also, of *Chel. cuneiceps*, instead of being pretty uniformly concave and even, as in *Chel. longiceps* and *Chel. caouanna*, is raised on each side between the middle line and the marginal alveolar plate into two convexities, as in *Chel. mydas* and *Chel. imbricata;* but the most prominent part of the palatal convexities (figs. 3 and 4, 21) is obtuse in *Chel. cuneiceps*, not sharp or angular, as in *Chel. mydas* and *Chel. imbricata*.

The palatal part of the vomer (13) forms the median longitudinal groove dividing the convexities, which are formed by the palatal processes of the maxillary bones. The small part of the alveolar border of the maxillary which is entire terminates in a sharp edge, extending about four and a half lines below the level of the palate.

The ridge of the palatines, which forms the anterior boundary of the posterior nostril, is not produced or bent below the level of the bony palate, as in *Chel. caouanna*, and as it is, although in a minor degree, in *Chel. mydas;* and there is not that concavity between it and the oblique palatal tuberosity which exists in the *Chel. mydas* and *Chel. imbricata*.

The pterygoids are more deeply (semicircularly) emarginate laterally than in any of the existing species of *Chelones*, and they are shorter in proportion to their breadth; they bound internally the lower apertures of the temporal fossæ, which are broader than they are long; in all the existing *Chelones* the opposite proportions prevail, and in *Chel. imbricata* especially the homologous apertures are twice as long as they are broad. The pterygoids, in the *Chel. cuneiceps*, develope a sharp ridge along their median suture; and short but well-defined processes at their anterior and outer angles. The channel or concavity upon the under part of the diverging portion of the pterygoid conducts obliquely into the temporal fossæ in the *Chel. mydas*; in *Chel. cuneiceps* it leads directly forwards upon the under surface of the anterior part of the pterygoids exclusively, as in the *Chel. imbricata* and *Chel. caouanna.*

In the *Chel. mydas* the malar approaches the mastoid very closely, and sometimes touches it by the posterior angle, thus separating the squamosal from the postfrontal; the extent of the union between the squamosal and postfrontal is also shorter in the *Chel. caouanna* than in the *Chel. imbricata.* In the extent of that union (between 12 and 27) the *Chel. cuneiceps* resembles the *Chel. imbricata*, as do likewise the *Chel. breviceps* and *Chel. longiceps.* But the *Chel. cuneiceps* differs from all the recent species in the form of the squamosal (27), which is bent upon itself, forming a slightly curved linear eminence, where the lower and smoother part of the bone is bent, and, as it were, pressed inwards towards the tympanic (28), against which it abuts. This modification is natural, not the effect of accidental pressure upon the fossil. The lower border of the malar (26), which intervenes between the maxillary and squamosal, is sharp but convex, as in *Chel. caouanna*, not concave as in *Chel. mydas*, nor nearly straight, as in *Chel. imbricata.* But the concave curve of the inferior margin of the squamosal (27) most resembles that in *Chel. imbricata.* The antero-posterior extent of the mastoid (8) is less proportionally than in any of the recent *Chelones*, and it forms a smaller share of the upper border of the large meatus auditorius. The articular part of the tympanic descends below the squamosal further than in the recent turtles; and its articular surface is more convex at its outer half, and more concave at its inner half; *Chel. imbricata* makes the nearest approach to the fossil in this respect. In the *Chel. mydas* and *Chel. caouanna* the articular surface is nearly flat.

As the supracranial scutella have left unusually deep and well-marked impressions on this fossil skull, I have reserved their description, and the comparison of their different forms and proportions in the several fossil species, to this place.

Three scutella occupy the median line of the upper surface of the cranium in the present species of *Chelone*, which, from the absence of any impression along the frontal and sagittal sutures, appear to have been single and symmetrical. The anterior and smallest answers to the " frontal" scute (*fr*); the next in size and position to the "sincipital" scute (*sy*); the hindmost and largest answers to the "occipital" scute (*oc*), which is usually divided, and forms a pair in existing *Chelones.*

The frontal scute is long, narrow, hexagonal, broadest across the antero-lateral angles, from which the impressions extend outwards to the supraorbital margin, which divide the "fronto-nasal" scute from the "supraorbital" scute (*ob*).

The sincipital scute is bounded on each side by a sigmoid curve, and both before and behind by an entering angle; it is broadest behind, and from the middle of the lateral border proceeds the transverse impression towards the back part of the orbit, which divides the " supraorbital" scute (*ob*) from the " parietal" scute (*pa*). The occipital scute is bounded laterally by straight lines, which slightly diverge as they extend backwards: there is no trace of an interoccipital scute. The parietal (*pa*) scute is the largest; impressions of five of its borders are preserved in the present fossil: the two exterior ones meet at an obtuse angle, a little above the middle of the *meatus auditorius externus*; the antero-external border uniting with the postorbital scute (*po*); the postero-external border with the external occipital scute (*eo*).

In the *Chelone breviceps* (T. I) the frontal scute is relatively larger than in the *Chelone cuneiceps*, and is nearly as broad as long. The sincipital scute is bounded laterally by two straight lines meeting at a very open angle, from which the transverse impression extends outwards between the supraorbital and parietal scutes. The straight lines bounding the sides of the occipital scute diverge from each other as they extend backwards more than they do in the *Chelone cuneiceps*.

In the *Chelone longiceps* (T. III) a still more different pattern of the supracranial scutation is presented. The occipital scutes (*oc*) are separated by an intervening interoccipital scute (*io*). The lateral borders of the sincipital scute are each bounded by three lines and two angles; the antero-lateral and postero-lateral angles being curved with the concavity outwards; and the transverse impression dividing the supraorbital scute (*ob*) from the parietal scute (*pa*), proceeds from the middle of the intervening straight border of the parietal. The frontal scute (*fr*) is long and narrow, broadest behind, with its lateral borders gradually converging to a point anteriorly; the impression dividing the supraorbital (*ob*) from the frontonasal scute (*fn*) proceeds from the middle of that lateral border. Neither the division between the frontal and sincipital, nor that between the sincipital and interoccipital scutes are well marked.

The *Chelone convexa* (T. VI, fig. 4), like the *Chelone longiceps*, has an interoccipital scute (*io*), and the sincipital scute (*sy*) has its sides bounded by three lines, of which the posterior one is curved with its concavity towards the occipital scute (*oc*), and so directed as to appear to form part of the posterior rather than the lateral border; the other two lines completing the lateral border and converging forwards, are divided or defined by a slight angle, from which the transverse impression proceeds outwards, which divides the supraorbital (*ob*) from the parietal (*pa*) scutes. The frontal scute (*fr*) is a small hexagon, relatively wider than in *Chel. longiceps* or *Chel. cuneiceps*. The impression dividing the supraorbital (*ob*) from the frontonasal (*fn*) scutes proceeds from the angle between the lateral and anterior sides of the frontal scute.

The *Chelone planimentum* (T. IX) is peculiar, and differs from all the foregoing species by the forward extension of the occipital scutes which join the supraorbital scutes, and thus divide the sincipital scute (*sy*) from the parietal scute (*pa*); the sincipital scute is correspondingly encroached upon, as it were, and narrowed, its broadest part being nearer the anterior end, at the angle between its two straight lateral borders, from which angle the impression extends outwards that divides the occipital from the supraorbital scute. The frontal scute (*fr*) is small and narrow, and the large supraorbital scutes meet in front of it at the middle line. They appear to be divided from the orbits by the encroachment of palpebral scutes (*pl*) upon the supraorbitary border. There appears to have been an interoccipital scute in the *Chel. planimentum*, as in the *Chel. longiceps* and *Chel. convexa*.

Amongst existing *Chelones* the interoccipital scute is constant only in the *Chel. caouanna*—the loggerhead of Catesby and Brown; but the sincipital scute in this species is vastly larger in proportion than in any of the fossils above described; and it is further distinguished by the peculiar division of the supraorbital and parietal scutes.

In the hawks-bill turtle (*Chel. imbricata*), the supracranial scutes leave as well-marked indentations upon the bones of the cranium as are seen in most of the fossil turtles, but the supraorbital scute is proportionably larger than in any of these, and the proportions and forms of all the other scutes are different. There are, also, two nasal scutes divided by a transverse groove from the frontonasals, which groove I have not yet met with in the corresponding part of any of the fossil Chelonian crania.

The skull of the *Chelone cuneiceps*, here described, is from the London clay of Sheppy.

CHELONE SUBCARINATA. *Bell*. Tab. VIII *A*.

The resemblance of this species to *Chelone subcristata* (p. 24, T. VIII) is so considerable, that it has not been without some hesitation that I have ventured to describe it as distinct. There are, however, certain characters by which it may be distinguished, and those of sufficient importance to be considered as specific. On comparing it with recent species, and even with most of the fossil ones from the same locality, there is a remarkable evenness in the arch of the carapace, which, with the exception of a slight carina on some of the posterior neural plates, to be hereafter mentioned, forms nearly a perfect arc of a circle, from the extremity of the costal plate of the one side to that of the other, without that flattening of the side which is seen in most other species.

The nuchal plate (T. VIII*A*, fig. 1, *ch*) has the posterior margin arched, and there is a short median process which goes to join the first neural plate (*s1*), in which respect it agrees with *Chel. longiceps* and with *Chel. subcristata*. This process is emarginate, to receive a slight triangular projection of the anterior margin of that plate. The first neural plate (*s1*) forms a parallelogram, the sides not being interrupted by any costal suture; the posterior suture of the first costal plate (*pl1*) extending to the second neural plate (*s2*). In this circumstance it differs from *Chel. subcristata, longiceps*, and *convexa*, and agrees with *Chel. breviceps*. This, however, may possibly be a variable character here, as it is in *Chel. longiceps;* in one specimen of which, now before us, Professor Owen found that the articulation in question was to the anterior part of the second, instead of the posterior part of the first, neural plate; in other words, that the first neural plate was the isolated one instead of the second. The remaining neural plates are hexagonal, becoming almost regularly shorter to the eighth; the lateral angles meeting the costal sutures being nearly at the same distance from the anterior margin in each, and in no one at all approaching a regular equilateral hexagon, as in many of the neural plates in *Chel. breviceps.* The first three, and the anterior half of the fourth neural plates are flat; but on the posterior half of the fourth commences a low carina, which becomes highest on the posterior half of the sixth ($s6$), and anterior half of the seventh ($s7$). It thus differs from *Chel. subcristata,* in which there is a distinct. short, sharp, longitudinal crest ($s1$) on the fifth and seventh neural plates, " and a similar crest is developed on the contiguous ends of the second and third neural plates." The ninth and tenth neural plates are wanting in the only specimen I have seen of the *Chel. subcarinata.*

The first costal plate is flat ($pl1$), but the remaining ones, to the seventh inclusive, are slightly hollowed along the middle, being raised towards the anterior and posterior margins, where they are articulated to the contiguous ones. The whole surface of the bones of the carapace is less smooth than in most other fossil species, and conspicuously less so than in *Chel. subcristata.*

In describing the forms of the vertebral scutes, ($v1$—$v4$), and of the costal ones as depending upon them, it is necessary, in order to arrive at any satisfactory comparison between these parts in different species, to bear in mind that a great change takes place in their outline during the growth of the animal; and that a vertebral scute, which, in a younger individual, has the middle of its outer margin exceedingly extended, so as to form a very acute angle, where the lateral margin of the costal scute joins it, and thus rendering it twice as broad as it is long, may in more advanced age have that angle very open, and having increased greatly in length, and scarcely at all in breadth from angle to angle, the length becomes greater than the breadth. Allowing, however, for this fact, there are doubtless considerable variations in this respect according to the different species, which are permanent and well marked. The first vertebral scute ($v1$) in the present species is quadrilateral, broader anteriorly; the second and third ($v2$, $v3$) hexagonal, with the outer margins slightly waved, somewhat broader in the middle at the angles than at the anterior and posterior margins, the comparative breadth at that part being rather greater than in the corresponding scutes of *Chel. subcristata,* and much less so than in *Chel. convexa, Chel. breviceps,* or *Chel. longiceps.* The fourth vertebral scute ($v4$) is also hexagonal, but the portion posterior to the lateral angles is narrowed and produced backwards. The last of the series is fan-shaped. The outline of the costal scutes follows of course that of the vertebral ones.

The plastron, in the specimen from which this description is taken (Pl. VIII, A, fig. 2), is more perfect than in that of almost any other fossil Chelonian I have seen. It agrees in its general form with that of *Chel. subcristata,* but is less extensive, as regards its bony surface, than in *Chel. longiceps* or even than in *Chel. breviceps.* The entosternal bone (*s*) is somewhat wedge-shaped, with the anterior margin triangular, and a short winged process on each side of the anterior third of the bone extending outwards and backwards. The posterior extremity of the bone, and the winged processes are dentate. The episternals (*es*) are aliform, tending backwards and outwards, and inclosing between them the head of the entosternal (*s*), and the anterior processes of the hyosternal bones (*hs*). The latter have the anterior processes extending forwards on each side of the entosternal, approximating at their extremity the aliform processes of that bone. The median or internal processes nearly meet on the median line, and the dentations are deep but slender; each hyposternal (*ps*) unites similarly with its fellow, and the posterior process extends backwards, in a long, narrow, triangular piece, uniting with the xiphisternal (*xs*), which latter forms a very elongated rhomb, the breadth of which is scarcely one fourth of its length, which in the present specimen is no less than two inches six lines. This form, with the elongation and narrowness of the posterior process of the hyposternal, gives to the hinder portion of the plastron in this species a narrower and more elongated outline than we find in almost any other; an approach to which is, however, indicated in the imperfect specimen of *Chel. subcristata* figured in Plate VIII.

The external notch, between the external process of the hyosternal and hyposternal, is deep and rounded. The central interspace is nearly quadrate, and about half as long again as it is broad.

	Inches.	Lines.
Length of the carapace as far as it is preserved . . .	9	5
Breadth of ditto from the extremity of the third costal plate on one side to that on the other	7	4
Ditto, following the convexity of the carapace	9	3
Length of plastron from the anterior margin of the episternal to the extremity of the xiphisternal	8	4
Breadth of ditto across the hyosternals	7	0

The only specimen of this species which I have seen is from Sheppy, and is in the fine collection of J. S. Bowerbank, Esq., F.R.S.

T. B.

SUPPLEMENTAL REMARKS

ON THE

TURTLES FROM THE LONDON CLAY AT HARWICH.

IN the progress of the works now carried on in a part of the Harwich cliffs, with a view to the acquisition of the remains of the animal tissues and bone-earth which form the nodules that are ground up and used as manure, many remains of the Chelonian reptiles which formerly frequented the seas from which those Eocene tertiary strata have been deposited have been discovered. Mr. Colchester, of Little Oakley, Essex, who carries on large works of this kind for the "Fossil Guano," as it is termed, has transmitted to me a number of the nodules in question. The most intelligible and instructive of these I have marked from 1 to 10 consecutively, and shall notice them here in the same order.

No. 1. *Chelone planimentum.* This is the half of an oval nodule of petrified clay, 20 inches in length, by 17 inches in breadth, exposing an irregular group of disarticulated bones of the carapace and other parts of the skeleton. The species is determined by a fragment of one of the costal plates with the connate rib. The plate measures $2\frac{1}{2}$ inches in breadth, the rib 8 lines, and forms the usual partial prominence from the even surface of the under part of the costal plate. Almost the whole of the very broad but short nuchal plate is recognisable: it measures 6 inches in transverse diameter, and only $1\frac{1}{2}$ inch in antero-posterior diameter. Part of the hyosternal bones, and the impression of the humerus are recognisable.

No. 2 is the half of a nodule, 20 inches in length and 17 inches in breadth, exposing part of the plastron, and some other bones of the skeleton of the *Chel. planimentum.* It shows well the natural form of the under and outer part of the hyposternal bone, which is much more deeply excavated than in the *Chel. crassicostata ;* the lower portion of the bone is narrower in proportion to its length, and the xiphisternals are also in proportion longer and narrower than in that species.

No. 3. *Chelone planimentum.* The half of an oval nodule, 17 inches in length and 13 inches in breadth. The fractured side exposing a cast of the inner surface of the carapace, which measures in length from the nuchal to the tenth neural plate inclusive $13\frac{1}{2}$ inches ; and in breadth, across the third pair of costal plates from one end of the projecting rib to that of the opposite side, 11 inches. The anterior contour of the carapace is well shown in this nodule, the marginal plates which join the nuchal plate being preserved. The free extremity of the rib attached to the third costal plate projects 1 inch 9 lines from that plate, and measures 7 lines in breadth, where it becomes free; the breadth of the plate being nearly 2 inches. The transverse curve of the carapace is shown by this specimen to be much less than in the *Chel. crassicostata*.

No. 4. *Chel. planimentum.* The nodule shows partly a cast of the outer surface of the carapace, with part of the carapace itself. The outer angles of the third and fourth vertebral scutes are here seen with the inner angle of the third costal scute. The outer angles of the vertebral scutes are more prominent than in *Chel. declivis, Chel. subcristata, Chel. subcarinata, Chel. convexa,* or *Chel. longiceps;* they resemble most those in *Chel. breviceps.* The breadth of the third costal scute is 4 inches. The characteristic angular ridge, formed by the narrow connate rib, where it projects from the lower surface of the costal plate, is well shown in this specimen.

No. 5. A nodule showing a cast of the under surface of the carapace seen from above, apparently of the *Chel. planimentum.*

No. 6. A nodule, 10 inches long by 9 inches broad, showing a still more imperfect cast of the under surface of the carapace, of apparently a younger specimen of the *Chel. planimentum.*

No. 7. A fragment of a nodule showing the outer dentated extremity of the left hyosternal of the *Chel. planimentum.*

No. 8. A portion of a nodule, with part of the carapace of the *Chel. planimentum,* showing the second to the seventh neural plates inclusive, and portions of the second to the seventh costal plates of the right side, with more or less of their bony substance broken away, exposing their coarse fibrous character, the fibres diverging on each side from the subjacent rib, as they extend obliquely towards the periphery of the carapace. The third neural plate is 2 inches 3 lines in length and 1 inch in breadth; it is crossed at its middle part by a moderately broad and deep channel, indicating the junction of the second with the third vertebral scute. The third neural plate is hexagonal; the two shortest sides being formed by the truncation of the contiguous angles of the second costal plates bending down a little to articulate with them. The fourth neural plate is 2 inches 6 lines in length, and 1 inch 4 lines across the broadest part. The anterior surface is concave, the posterior convex; the two longest sides converge towards the posterior surface, and are straight. The fifth and sixth neural plates progressively decrease in length, without a proportionate decrease in breadth. The breadth of the fourth costal plate is 2 inches 3 lines at its peripheral extremity: its length is 6 inches; the rib projects 2 inches beyond it. The upper

6

surface of the neural and costal plates is so minutely fibrous or striated as to seem at first sight almost smooth. The upper surface of the costal plate seems naturally to be slightly concave in the direction of the axis of the carapace, but not so much as in *Chel. crassicostata*, and the rib is much bent lengthwise.

No. 9. *Chelone crassicostata* (T. XIII*A*). This instructive specimen is contained in a subspherical nodule, 13 inches long by 12 inches broad, exposing a large proportion of the outer surface of the carapace, with more than one half of the circle formed by the marginal plates (*m*7—*py*). The carapace has been fractured, and the ribs of the left side dislocated and pressed down below those of the right. The third (*pl*3) to the eighth (*pl*8) costal plates inclusive are present on the left side; the fifth to the eighth on the right side, and the neural plates from the fourth to the pygal plate (*py*) inclusive. The fourth, fifth, and sixth neural plates are hexagonal, with the anterolateral sides shortest, and chiefly remarkable for their great breadth in proportion to their length. The seventh and eighth are small, and more regularly hexagonal. The ninth is a broad sub-crescentic plate, with the broad concave side backwards, and the space between this and the pygal plate is filled up by an equally broad but pentagonal neural plate. The length of the ninth and tenth neural plates, with the pygal plate inclusive, is 2 inches 9 lines. The pygal plate is subquadrangular and broadest behind, where it is slightly emarginate. The length of the fourth to the eighth neural plate inclusive is 3 inches 8 lines. The upper surface of the bones of the carapace is almost smooth. That of the costal plates is chiefly remarkable for its concavity transversely, or in the direction of the axis of the carapace, which is to a greater degree than in the *Chel. subcristata* or *Chel. longiceps*; the lines of the sutural union of these plates with each other forming so many ridges across the sides of the carapace. The degree of curvature or convexity in the direction of the length of the costal plate is much greater than in the *Chel. planimentum*. The length of the third costal plate is 3½ inches, its breadth at the outer extremity, 1 inch 4 lines; the breadth of the rib where it projects beyond it is 9 lines. The margin of the plate attached to that rib is 1 inch 4 lines in length, and 8 inches in breadth. The margin of the plates gradually increases in breadth towards the posterior part of the carapace, the one joining the pygal plate being 1 inch 2 lines in breadth. The general form of the carapace of the *Chel. crassicostata* is shown by the present specimen to have been that of a full oval, with a gently festooned border, not pointed behind.

No. 10. *Chelone crassicostata* (T. XIII*B*.) A still more remarkable example of this species was kindly transmitted to me by the Rev. S. N. Bull, M.A., of Harwich, of which a figure is given in T. XIII*B*. When it first came into my hands it was an unpromising semioval nodule, 10 inches in length by 7 inches in breadth, presenting on its convex surface portions of the posterior neural and costal plates, with their external surface entire; but no trace of plastron on the flattened side. The degree of convexity formed by the costal plates equalled that of the most dome-shaped tortoise. The flatter surface of the nodule was slightly convex, which I thought might arise from a layer of petrified clay adhering to the plastron. A portion of the cranium was indicated at the produced angle of the nodule. To ascertain whether this remarkable degree of convexity of the carapace, both lengthwise and transversely, was natural, I had the matrix carefully removed, with the permission of the owner of the specimen, and the same was done on the opposite side, with a view to expose the plastron. Instead of finding a plane plastron where it was expected, in its natural horizontal position, it was found to have been crushed inwards, as represented in fig. 2, by the pressure of a hard petrified mass as big as a paving-stone, which had been forced in upon this part of the body of the turtle whilst in a decomposing state; and when finally lodged in the clay, the carapace and plastron, as they became dislocated, had become more or less moulded upon it; and thus was produced the convexity which originally attracted my attention. In the breadth of the connate rib, as compared with that of the costal plate, in the extent of the free extremity of the rib, in the degree of concavity of the upper surface of the costal plate and the curvature lengthwise, the distinctive characters of the *Chel. crassicostata* are well shown. The same characters are likewise presented by the parts of the plastron, as in the breadth of the xiphisternals (*xs*), the curvature of the hyosternal (*hs*), and the form of the coracoid. The two scapulæ, with the connate acromial clavicles, are preserved, with the head of one of the humeri. A part of the basis cranii, showing the broad diverging pterygoid, with their characteristically-channeled inferior surface, is shown in fig. 2; these grooves are not so deep, however, as in *Chel. longiceps*, but are more like those in *Chel. cuneiceps*.

I beg to record my obligations to Mr. Bowerbank for the suggestion, and to Mr. Bull for his ready response to it, to which I owe the opportunity of examining this specimen of the thick-ribbed turtle of the Harwich cliffs.

A fossil mandible of a Chelonian, in the collection of the Marchioness of Hastings (figured, of the natural size, in T. XIX*D*, figs. 1 and 2), most resembles that part in the genus *Chelone* by its general form and proportions, and especially by the configuration of the biting and grinding surface of the jaw (fig. 2). The symphysis is confluent; convex in both directions below; longer than in the *Chel. mydas* and the *Chel. breviceps* of Sheppy (T. I, fig. 3, 32); but not so long as in the turtles from Harwich (T. IX, fig. 1, and T. XI, fig. 3) and Bracklesham, or as in the *Chelone longiceps* of Sheppy. The rami diverge more from each other than in the lower jaw of the *Chel. convexa* (T. VII, fig. 3) of Sheppy.

A ridge, commencing at the fore part of the upper surface of the symphysis, passes backwards, and divides the two ridges, diverging and circumscribing with the outer sharp margins of the jaw an elliptical concave space on each side; the space between the diverging ridges is raised and rough: this part has been fractured. In the *Trionyx*, of which genus so many fine examples have been met with at Hordwell, the upper part of the symphysis presents an uniform concavity; and this part of the jaw is narrower and more produced. I have not yet seen the mandible of any Emydian or land-tortoise resembling the present fossil so closely as some of the marine species above cited. A large species of *Emys* has, however, left its remains in the same deposits at Hordwell as the *Trionyces* next to be described.

A retrospect of the facts above detailed, relative to the fossil Chelonians of the genus *Chelone*, or marine family of the order, leads to conclusions of much greater interest than the previous opinions respecting the Chelonites of the London clay could have suggested. Whilst these fossils were supposed to have belonged to a fresh-water genus, the difference between the present fauna and that of the Eocene period, in reference to the Chelonian order, was not very great; since the *Emys* or *Cistuda Europœa* still abounds on the continent after which it is named, and lives long in our own island in suitable localities. But the case assumes a very different aspect when we come to the conviction that the majority of the Eocene Chelonites belong to the true marine genus *Chelone*; and that the number of species of these extinct turtles already obtained from so limited a space as the Isle of Sheppy, exceeds that of the species of *Chelone* now known to exist throughout the globe.

Notwithstanding the assiduous search of naturalists, and the attractions to the commercial voyager which the shell and the flesh of the turtles offer, all the tropical seas of the world have hitherto yielded no more than five* well-defined species of *Chelone*; and of these only two, as the *Chel. mydas* and *Chel. caouanna*, are known to frequent the same locality.

It is obvious, therefore, that the ancient ocean of the Eocene epoch was much less sparingly inhabited by turtles; and that these presented a greater variety of specific modifications than are known in the seas of the warmer latitudes of the present day.

The indications which the English eocene turtles, in conjunction with other organic remains from the same formation, afford of the warmer climate of the latitude in which they lived, as compared with that which prevails there in the present day, accord with those which all the organic remains of the oldest tertiary deposits have hitherto yielded in reference to this interesting point.

That abundance of food must have been produced under such influences cannot, of course, be doubted; and we may infer that, to some of the extinct species, which, like the *Chel. longiceps* and *Chel. planimentum*, exhibit either a form of head well adapted for penetrating the soil, or with modifications that indicate an affinity to the *Trionyces*, was assigned the task of checking the undue increase of the now extinct crocodiles and gavials of the same epoch and locality, by devouring their eggs or their young; becoming probably, in return, themselves an occasional prey to the older individuals of the same carnivorous Saurians.

* Mr. Gray, for example, includes the *Chelone virgata* and *Chelone maculosa* of Dumeril and Bibron as varieties of the *Chelone mydas*.

Family—FLUVIALIA.

Genus—TRIONYX.

THE Chelonian Reptiles called " Soft Tortoises," forming the genus *Trionyx* of Geoffroy St. Hilaire,* and the family *Fluvialia* seu *Potamites* of MM. Duméril and Bibron,† resemble those of the genus *Chelone* (family *Marina* seu *Thalassites*, Dum. and Bibr.) in the extremity of the vertebral rib, or pleurapophysis, projecting freely from below the end of the connate costal plate,‡ and in having the plastron incompletely ossified; but they are characterised by the still more incomplete ossification of the margin of the carapace, which retains much of its primitive soft, cartilaginous state; and they are further distinguished by the reduced number of the toes—three on each foot—which are armed with claws, the other two toes serving to support a swimming web; the name of the genus has reference to this peculiarity.

The head is depressed, elongated, and, in the recent animal, the nostrils are prolonged into a short tube, terminated by a small fleshy appendage like an elephant's proboscis. The outer surface of the dermal bones of the carapace, and of the corresponding parts of the plastron, is variously sculptured, usually by sinuous grooves and rugosities, as if wormeaten; and to such a degree in some species, as to give the parts a tuberculate character. The cuticle is soft and flexible, not developed into scutes; and there are accordingly no impressions like those that indicate the presence of the " tortoise-shell" plates in the skeleton of the existing turtles and in the petrified plastrons and carapaces of the extinct species of the marine family.

" Hitherto," write the meritorious authors of the elaborate ' Erpételogie Générale,' one has not observed any species of this family (*Potamites*) in our European rivers; all those which have been described, and of which the habitat is known, have come from the streams, rivers, or great fresh-water lakes of the warmer regions of the globe." (Tom. ii, p. 469.) The beautifully-preserved evidences of the species about to be described, which have chiefly been obtained by the Marchioness of Hastings from one limited locality, attest the abundance of the *Trionyces* in the fresh-waters of our latitudes during the Eocene period of geology.

The characters by which MM. Wagler and Duméril have divided the species of

---

* Annales du Muséum d'Histoire Naturelle, tom. xiv.

† Erpétologie Générale, 8vo, tom. ii, p. 461.

‡ This character is well exemplified in the Marchioness of Hastings's unique and beautiful specimen of the *Trionyx rivosus*, T. XVIII*A*.

Trionyx, Geoffr., into two genera, are not such as can be decisively recognised in the fossil carapace. A difference of convexity of that part by which the *"Cryptopodes"* are said to differ from the *Gymnopodes*, is not one that the comparative anatomist and palæontologist would recognise as valid for the distinction proposed.

Upon the whole, the fossil specimens in which that character can be compared, agree rather with the *Gymnopodes* of Dum. and Bibr., but with a range of diversity which is exemplified by Tab. XVI*A*, and XVII. So much of the plastron as I have been able to compare, agrees likewise with the bones of that part in the *Gymnopodes*, but, in the absence of more certain characters, and with doubts as to the necessity or desirableness of the subdivision proposed for the recent species, I shall retain the name *Trionyx* for all the fossils that manifest, in their petrified remains, the characters of the Geoffroyan genus.

In the second part of my 'Report on British Fossil Reptiles' I showed that certain fossils of the Wealden formation and of the Caithness slate (new red sandstone) had been referred erroneously to the genus *Trionyx*, and that the only unequivocal remains of that genus which had been seen by me at that period (1841) were from Eocene deposits at Sheppy, Bracklesham, and the Isle of Wight, in which latter locality they were associated, as in the Paris basin, with remains of the *Anoplotherium* and *Palæotherium*.

I have since had the opportunity of examining fossil specimens of *Trionyx* from other localities, but always, however, from formations of the Eocene period, and I shall commence their description with one of the most perfect and beautiful examples of these Chelonites, which was obtained by the Marchioness of Hastings from the Eocene sand of the Hordwell Cliff, Hants.

TRIONYX HENRICI. *Owen.* Tab. XVI.
Report of the Seventeenth Meeting of the British Association, 1847, p. 65.

Although the characteristics of the genus are readily recognisable in fossil fragments of the carapace and plastron, from their comparative flatness and the sculpturing of the outer surface, the species of *Trionyx* are with difficulty determinable, if at all, from such specimens; and it is usually necessary to have a considerable part of the carapace, in order to ascertain its composition, contour, and degree of convexity. Some species, indeed, e. g. *Trionyx rivosus* (T. XVIII*A*), *Trionyx marginatus* (T. XIX*), together with the *Trionyx spinosus** and *Trionyx sulcatus* of Kutorga, would seem to be characterised by particular patterns of the irregular surface of the bones of the carapace, which character, therefore, a fragment may suffice to manifest; but this is not the case with the ordinary rugose and vermiculate species. Cuvier accordingly

* This is quite a distinct species from the *Trionyx spiniferus* of Lesueur.

admits, with respect to the portions of *Trionyx* found abundantly in the gypsum of the environs of Paris, associated with the Palæotheres, Anoplotheres, and other extinct animals of the Eocene epoch, that he could find nothing in those fragments to authorize him to fix their specific characters.* The compilers of the labours of palæontologists have, as usual, been affected by no such scruples, and have not hesitated to assume a knowledge, which Cuvier did not feel himself entitled to claim, viz. that of the fact of the specific distinction of the *Trionyx* of the Montmartre quarries : but I do not find that they have added anything to its history except the name of *Tri. Parisiensis.* It is probable, from the analogy of our own Eocene deposits, that more than one species of *Trionyx* may have left its remains in the Parisian localities of the corresponding geological formation.

The fossil remains of *Trionyx* from the tertiary deposits of the Gironde,† Lot-et-Garonne,‡ Montpellier, and Avary, were not sufficiently characteristic to permit the great anatomist and founder of Palæontology to infer more than the existence of the particular genus of fresh-water Chelonia in question in those formations. The only specimen of fossil *Trionyx* in which Cuvier recognised characters distinguishing it from the known existing species, is that which M⁰. Bourdet first described§ under the name of *Trionyx Maunoir*, from the Eocene quarries at Aix. Cuvier has given reduced views of a large proportion of its carapace and half its plastron in the ' Ossemens Fossiles,' tom. v, pt. ii, Pl. XV, figs. 1 and 2. This description and the figure of the carapace serve to elucidate by comparison the characters of the more perfect specimen of *Trionyx* here described from the Eocene of Hordwell.

In the first place, the contour and proportions of the entire carapace of the *Tri. Henrici* differ from those of the *Tri. Maunoir*. The carapace of the *Tri. Henrici*, which is formed, as usual, by the neural plates (s_1, s_2, &c.) and eight pairs of costal plates (pl_1—8), measures 10 inches 8 lines in length, and 11 inches 2 lines in breadth, in a straight line across the third (pl_3) costal plate, where it is widest. In *Tri. Maunoir*, the neural plates (plaques vertébrales) rise a little above the plane of the carapace, as in the *Tri. ferox* (*Tri. carinatus*, Geoffr.‖) : in the *Tri. Henrici* there is no trace of this carinate structure ; the neural plates arc flat, and on a level with the broad costal plates articulated with them ; in which characters it resembles the *Tri. gangeticus*, Cuv., and *Tri. javanicus*, Cuv.

The first costal plate (pl_1) is broader than it is long in *Tri. Maunoir*; in *Tri. Henrici* its breadth is little more than half its length, and decreases as it recedes from

---

* "Mais je n'ai rien trouvé dans ses débris qui m'autorisât à en fixer les caractères spécifiques." (Ossemens Fossiles, tom. v, pt. ii, p. 223.)

† Cuvier, tom. cit., pp. 225, 227.

‡ The skull of the *Trionyx* from this locality showed a slightly different profile from that of any of the existing species.

§ Bulletin de la Société Philomathique, 1821.

‖ Annales du Muséum, tom. xiv, pl. 4, 1809.

the neural plate. The second costal plate, on the contrary, is broader at its lateral than at its mesial end in *Tri. Henrici*, whilst its breadth is equal at both ends in the figure given by Cuvier of the *Tri. Maunoir*. The thickness of the costal plates, in proportion to their breadth, is shown in T. XIX*B*, figs. 4 and 5; the degree of projection of the connate rib from the inner surface of the costal plate is given in figure 6. The peripheral border of the carapace is not grooved in this species, as in the *Tri. circumsulcatus*, fig. 3.

The degree of transverse convexity of the carapace of the *Tri. Henrici* is the same as that of the *Tri. Ægyptiacus*, and as that attributed to the *Tri. Maunoir*.*

The nuchal plate is wanting in Lady Hastings's specimen; the one which is figured in T. XVI, fig. 3, is from the same locality at Hordwell, but does not belong to the carapace, fig. 1, although it has probably belonged to one of the same species, from the contour of its hinder border.

The first neural plate ($s1$) does not project beyond the adjoining anterior borders of the first costal plates ($pl1$) as it does in *Tri. subplanus*, *Tri. ferox*, and *Tri. javanicus;* nor do those borders, as they recede from the neural plate, curve forwards beyond it, as in *Tri. javanicus*[†] and *Tri. corômandelicus*.[‡]

The anterior border of *Tri. Henrici* is slightly concave and gently undulated, as in the *Tri. Ægyptiacus*, and is also rough and sutural, showing that the anterior azygos or nuchal plate (" pièce impaire," Cuv.) had been immediately articulated with it, as it is in *Tri. Ægyptiacus*.[§]

The fossil specimen of the nuchal plate, figured in T. XVI, fig. 3, shows, by the sutural structure of its posterior border, that it articulated with the anterior sutural border of the carapace to which it belonged, and which, as already remarked, belonged probably to the species *Tri. Henrici*, though not to the individual the carapace of which is figured in T. XVI, fig. 1.

The neural plate ($n1$) is longer in proportion to its breadth, and the corresponding costal plates ($pl1$) are narrower at their extremities than in *Tri. Ægyptiacus*. The second costal plates ($n2$) are broader at their extremities than in *Tri. Ægyptiacus;* they resemble those in *Tri. subplanus*.[||]

The first four neural plates in *Tri. Henrici* slightly expand posteriorly, and have their posterior angles cut off; the fifth ($n5$) is a narrow plate with entire angles; the sixth ($n6$) is expanded anteriorly, and has its anterior angles cut off; the seventh ($n7$) has also its anterior angles cut off, but is rounded behind, and, as it were, obliterated by the extension of ossification from the costal plates into the dermal cartilage above

---

* " Sa convexite transversale est telle, que la flèche de l'arc est moindre du cinquième de la corde." (Cuvier, Ossemens Fossiles, tom. v, pt. 2, p. 223.)

† Annales du Muséum, tom. xiv, pl. 3, *A*.

‡ Ibid., pl. 5, fig. 1.

§ Ibid., pl. 2, A, *a*.

|| Ibid., pl. 5, fig. 2.

the neural spines. The eighth neural plate is wholly obliterated or superseded by a similar encroachment and union of the eighth pair of costal plates ($pl8$). Almost the same modification is represented by Geoffroy in the carapace of the *Tri. Ægyptiacus*;[*] but the general proportions of the carapace of the *Tri. Henrici* are more like those in the *Tri. subplanus,* in which the eighth neural plate exists in the interspace of the eighth pair of costal plates, as it does likewise in *Tri. Mannoir.*

All the exterior surface of the expanded parts of the neural spines and ribs is roughened or sculptured with a moderately fine vermicular pattern, the undulatory grooves having a tendency to a concentric arrangement at the peripheral surface of the carapace, and in general passing uninterruptedly from one costal plate to another : the pattern is effaced from about one third of an inch of the border of the carapace, which presents a surface like that of a coarsely-woven cloth. The extreme border is rather suddenly bevelled or rounded off from above downwards, and is thinner than the border of the costal plates that articulates with the neural plates. The natural extent of the ordinary narrow extremities of the ribs cannot be determined from the present specimen of the *Tri. Henrici* ; they form the usual slight relief along the middle of the smooth under surface of the connate costal plates ; and do not subside at any part of their course to the level of the under or inner surface of the plate.

T. XVI, fig. 1, shows the upper surface of the carapace of the *Tri. Henrici,* half the natural size.

Fig. 2, in outline below, gives the curve and degree of transverse convexity across the middle of the carapace.

Fig. 3, the nuchal plate of apparently the same species of *Trionyx,* half the natural size.

T. XIXB, fig. 4, shows the outside view of the third costal plate, right side, natural size ; fig. 5, sutural border of the same plate, showing its thickness and the degree of curvature ; fig. 6, the peripheral border of the same plate with the connate rib.

All these specimens were discovered by the Marchioness of Hastings in the Eocene sand at Hordwell, and are preserved in her ladyship's Museum at Efford House, near Lymington, Hampshire. The species is dedicated to her ladyship's husband, Captain Henry, R.N.

In the figure of the carapace of the *Trionyx* (*Tri. subplanus*) in Cuvier's pl. xiii, fig. 5, ' Ossemens Fossiles,' tom. v, pt. ii, the costal plates do not bear the same numbers as the corresponding neural plates ; the anterior costal plate is marked $a1$, whilst the corresponding neural plate is $b2$; the rib or pleurapophysis of the first dorsal vertebra, which is marked $c1$, is short, and is applied to the under and fore part of the second rib which supports the first costal plate. In T. XVI, the dermal ossifications of the carapace bear the same letters and numbers as the homologous parts in the previous plates, and in the woodcut, fig. 1, p. 3.

---

[*] Ibid., pl. 2, *A.*

Trionyx Barbaræ. *Owen*. Tab. XVI*A*.

This species, like the *Trionyx Henrici*, is most satisfactorily and beautifully represented by an entire carapace in the collection of the Marchioness of Hastings, to whose indefatigable researches in the locality of the Eocene sand at Hordwell Cliff, its discovery is due, and by whose skill, tact, and patience it has been faithfully restored from its original fragmentary state.

The carapace is more slender in proportion to its length, and deeper or more convex in proportion to its breadth, than in the *Tri. Henrici*. In this species, as is shown in T. XVI, the breadth is greatest towards the fore part of the trunk; in the *Tri. Barbaræ* this is the narrower part, and increases in breadth towards the middle of the carapace (*pl*4).

The antero-posterior diameter or length of the nuchal plate is greater in proportion to its transverse diameter or breadth, and the arched ridge on its inner surface is less strongly developed than in *Tri. Henrici*. On the outer surface the smooth anterior border, where the plate would seem as if cut away obliquely to an edge, is more extensive in comparison with the rough, worm-eaten surface in the *Tri. Henrici* (T. XVI, fig. 3), or those in the nuchal plate of *Tri. incrassatus*, T. XVIII, fig. 1, *ch*. The median part of the anterior border is more deeply excavated, and the lateral borders less deeply dentated in the *Tri. Barbaræ*.

The whole of the posterior border of the nuchal plate is thick, sutural, and is articulated to the first neural plate and the anterior costal plates (*pl*1); the middle part extending backwards to unite with the neural plate, by which also *Tri. Barbaræ* differs from *Tri. Henrici*.

The first neural plate is shorter and broader in proportion to the length of the costal plates than in the *Tri. Henrici*, but presents a similar shape, the sides being parallel, and the posterior angles truncate; in the three succeeding neural plates the sides converge towards the anterior end, but the posterior angles continue to be cut off. The fifth neural plate is oblong and quadrangular, as in *Tri. Henrici*, T. XVI. In the sixth neural plate the fore part is the broadest, and its angles are truncate; the seventh is a subtriangular and not fully-developed plate; the corresponding pair of costal plates meeting behind it. The eighth pair of costal plates (*pl*8) similarly supersede and take the place of *s*8, by meeting and joining at the middle line, but the left is the broadest, not the right.

The first costal plate (*pl*1) is longer in proportion to its breadth (or antero-posterior diameter), which is also more equally preserved throughout its length than in the *Tri. Henrici*, and the connate smooth rib is less prominent on its under surface. The inner and anterior angles of this surface do not show the depression formed by the head of the vertical scapula, which is present in that part of the stronger *Tri. Henrici*.

A well-marked distinctive character is also afforded by the seventh costal plate (*pl7*), from which the free end of the connate rib projects at the anterior angle of the dilated end in *Tri. Barbaræ*, and the free border of that end describes a straight line transverse to the axis of the carapace.

The free borders of the eighth pair of costal plates are on the same transverse line, and the posterior part of the carapace is censequently truncate and straight.

The lateral margin of the carapace is more gradually bevelled down, and to a less obtuse edge than in the *Tri. Henrici*.

The length of the carapace of the *Tri. Barbaræ*, from the fore part of the first neural plate to the hind border, is nine inches and a half; the greatest breadth of the carapace, in a straight line across the fourth pair of costal plates, is nine inches ten lines. The total length of the carapace is eleven inches and a half.

The free end of the connate rib projects entire from the fifth, sixth, and seventh costal plates.

The character of the sculpturing of the outer surface of the costal plates is very similar to that in the *Tri. Henrici*: the tendency to the concentric arrangement of the raised lines is equally well marked in *Tri. Barbaræ*, and is accurately given in Mr. Erxleben's beautiful plate.

The carapace is not only more arched transversely, but it differs from that of *Tri. Henrici* in being slightly depressed along the middle line, as is indicated in fig. 2, T. XVI*A*.

This beautiful species of *Trionyx* is dedicated, with much respect, to its accomplished discoverer, Barbara, Marchioness of Hastings, and Baroness Grey de Ruthyn.

TRIONYX INCRASSATUS. *Owen.* Tab. XVII, XVIII, and XIX.

This species of *Trionyx*, from Eocene formations of the Isle of Wight, resembles in general form the *Tri. Henrici* of the Hordwell sand, but differs from it in the anterior internal angle of the first costal plate (*pl1*, T. XVII and XVIII) being cut off, like that of the second and succeeding costal plates : it also differs in the greater length of the second costal plate as compared with the breadth of its outer end, and in the greater breadth of the outer end of the sixth costal plate (*pl6*, T. XVII), the outer or terminal border of which is more convex. The nuchal plate (*ch*, T. XVIII) articulates with the whole anterior border of the first neural (*s1*) and costal plates (*pl1*), but sends backwards a process from near the middle of its posterior border, which fits into the space left between the truncated antero-internal angles of the first costal plates and the first neural plate. In this respect it resembles the nuchal plate of *Tri. Barbaræ* (T. XVI*A*), but the difference of general shape between this more delicately formed species, and the one under consideration, is well marked, and decisive as to their specific distinction. The anterior border of the nuchal plate of *Tri. incrassatus* is smooth, slightly channeled, and feebly emarginate at the middle part; the plate sends out three short, tooth-like processes on each side; the posterior angle forms a fourth process which articulates with the true costal part, or end of the second rib, connate with the first costal plate (pl_1). The first neural plate (s_1, T. XVIII) is rather broader in proportion to its length than in the *Tri. Henrici*. The second (s_2) and third (s_3, T. XVII) do not expand so much behind; the vermicular pattern is broken into distinct tubercles upon these plates. The posterior lateral sides of the hexagonal neural plates are relatively longer than in those of *Tri. Henrici*. The fifth neural plate (s_5, T. XVII) extends backwards beyond the fifth pair of costal plates (pl_5, compare with T. XVI) and articulates with the sixth pair of costal plates; but the eighth and part of the seventh neural plates are superseded by ossification, extending from the seventh and eighth pairs of costal plates to the median line, where those plates articulate with each other, as in the *Tri. Henrici* and *Tri. Barbaræ*. The inner surface of the nuchal plate (*ch*, fig. 2, T. XVIII) is divided by a transverse, slightly interrupted ridge, gently concave backwards, into two nearly equal parts; the posterior one being most excavated. The inner surface of the first costal plate (pl_1, T. XVII and XVIII) presents the prominence (c_2) left by the fracture of the vertebral end of the second rib, where it becomes connate with that plate, and also the oblique ridge (c_1) formed by the attachment of the expanded end of the first short rib. The free end of the second rib (c_2) is short, obtuse, depressed, convex above and flat below; the body of this rib has subsided to the level of the inner smooth surface of the costal plate, with which it has become completely blended. A small portion of the body of the second vertebra is preserved in connexion with the long neural arch, showing that it was slightly carinate at the under surface. The breadth of the third rib (c_3), where it becomes connate with the second costal plate (pl_2), is rather more than one third the breadth of that part of the plate; the rib at first sinks almost to the level of the under surface of the plate, and then gradually rises, increasing in breadth to its free extremity. The true pleurapophysial portions of the succeeding costal plates (4, 5, 6, 7, 8, and 9, T. XVII) are better defined by outline grooves, but their degree of prominence is slight, except in the last pair (9), which have been liberated from the superincumbent costal plates (pl_8) before they reached their posterior borders.

The minute accuracy and beauty of Mr. Erxleben's lithographs supersede the necessity of further verbal description of these rare and singularly well-preserved fossils.

T. XVII gives an inside view of the almost entire carapace of the *Tri. incrassatus;* and T. XVIII gives an outside (fig. 1) and an inside view (fig. 2) of the fore part of the carapace of the largest individual of the same species of *Trionyx*, from the Isle of Wight, showing the nuchal plate (*ch*) in its natural articulation with the anterior neural and costal plates.

One character by which these carapaces differ from those of the *Tri. Henrici* or *Tri. Barbaræ* is the abrupt, almost vertical, border of the carapace, which is formed by the peripheral ends of the costal plates: these increasing in thickness as they approach that end, render the border characteristically thick: the specific name—*incrassatus*—has reference to this structure. The border is not grooved, and it is slightly produced above the projecting end of the subjacent rib, where it slopes a little down to the connate rib (T. XVIII, fig. 1). This structure will serve to distinguish a detached costal plate of the *Tri. incrassatus* from one of the *Tri. circumsulcatus* (T. XIX*B*, figs. 1, 2, 3) ; and the verticality and thickness of the margin will equally distinguish it from one of the *Tri. Henrici* or *Tri. Barbaræ*.

The chief value of the specimen (figured in T. XIX) is derived from the fact, that several other bones of the same skeleton were discovered with it; and these I next proceed to describe.

T. XIX, fig. 1, is the entosternal piece of the plastron, having the characteristic form of the chevron ; it is broadest and most compressed at the median junction of the two crura, which increase in thickness and diminish in breadth as they diverge. The branches are relatively more slender than in the *Tri. Ægyptiacus** and *Tri. Javanicus ;*† they resemble those of the *Tri. carinatus*‡ and *Tri. gangeticus.*§.

Fig. 2 2′ is the lower branch of the left episternal : it is slender, gradually tapering to a point, flattened above or on the inner surface, convex behind, grooved along the margin next the entosternal. This piece, in its length and slenderness, resembles the corresponding part in the *Tri. carinatus* and *Tri. gangeticus.*

Fig. 3 3′ is the left hyposternal and part of the left hyosternal; the latter (*hs*) includes the mesial border, showing the relative extent of the angular part that sends off the ridged tooth-like processes, which are two in number, the anterior one notched or subdivided. The exterior, connate, rough, and tuberculate dermal plate stops at the base of these processes. The hyposternal (*ps*) has the nearest resemblance to that of the *Tri. gangeticus* figured by Cuvier,‖ but differs by the number of short toothed processes from its median and inferior border, and by the more slender base supporting the two long, lateral, striated, pointed processes. The tuberculate dermal plate covers all the exterior of the hyposternal to the roots of the pointed processes. The notch for the reception of the xiphisternal is rounded at the bottom.

Fig. 4 shows the long, rib-shaped, but straight scapula (51); its head forms two thirds of the glenoid cavity for the humerus; the body, flattened behind, convex in front, gradually contracts as it ascends, and terminates in an obtuse point; the

---

* Geoffroy, loc. cit., pl. 2, fig. B, *o.*

† Ibid., pl. 3, fig. B, *o.*

‡ Ibid., pl. 4, fig. B, *o.*

§ Cuvier, Ossemens Fossiles, tom. v, pt. ii, pl. 12, fig. 46.

‖ Loc. cit.

clavicular process (58) is shorter than the scapula, and slightly expands at its extremity. Both parts are longer and more slender than the homologous ones of the recent *Trionyx* figured by Cuvier.*

Fig. 5. The coracoid has the expanded, slightly curved form characteristic of the genus; it is not so broad as that figured by Cuvier.†

Fig. 6 is the iliac bone, short, thick, curved, subcompressed, attenuated and striated at its sacral extremity; the enlarged articular end is divided into three facets: two oblong and rough for sutural junction with the ischium and pubis; one smooth, and the smallest of the three, for the acetabulum.

Fig. 7 is the almost entire right femur; its convex, long oval head, bends inwards from between the two trochanters, of which the external and largest is broken off. The shaft bends backwards, and gradually expands to the feebly divided convex condyles. All the characteristics of the modifications of the femur in the *Trionyx* are here preserved.

Fig. 8 is a claw-bone, natural size.

Fig. 9 9′ 9″ are three views of the sixth cervical vertebra of the same *Trionyx*. This may be recognised by the broad, depressed, posterior surface of the centrum, partially divided into two cavities, side by side (c'); the seventh cervical has the two cavities there quite separated from each other; the fifth and preceding cervicals have the posterior surface of the centrum with a single cavity; so that the sixth cervical is the only one which has a single convexity in front (c, fig. 9′), and a double concavity (c', fig. 9″) behind. The body is long, slender, compressed in the middle, with one median inferior ridge anteriorly, and a pair of inferior ridges posteriorly ending in hypophysial tuberosities (fig. 9, yy), which support, as it were, the posterior articular cups. A short, obtuse diapophysis projects from each side of the fore part of the centrum. The prezygapophyses (z) support slightly convex, oblong, articular surfaces; the zygapophyses (z') are long, diverge, and support concave, oblong surfaces looking downwards. There is no spine; the neural arch is complete above the middle third of the centrum, the canal expanding towards both its wide, oblique outlets; this modification of course relates to the great extent of motion between contiguous vertebræ, and the necessity for providing against compression of the myelon during their rapid inflections and extensions.

The specimens of *Tri. incrassatus* here described are preserved in the Museum of the Marchioness of Hastings, by whose kind and liberal permission they, with other rare Chelonites, have been described and figured for the present Monograph.

---

* Loc. cit., pl. 12, fig. 4.
† Loc. cit., pl. 12, fig. 4.

TRIONYX MARGINATUS.   *Owen.* Tab. XIX+.

A more obvious character than that pointed out at the peripheral border of the costal plate in *Trionyx incrassatus,* and serving better and more readily to determine such elements of the carapace, is the ridge with minute parallel striæ, which extends along the upper surface, close to the anterior and posterior borders of the costal plates in the species of *Trionyx* which I have on that account distinguished by the specific name of *marginatus.*

Mr. Erxleben has well given this character in the reduced view of the carapace (T. XIX+).

The border-pattern gradually becomes narrower and fades away before it reaches the outer end of the costal plates; it is also wanting on the anterior border of the first costal plate (*pl*1), and on the posterior border of the last (*pl*8).

The outer ends of the costal plates, which constitute the greater portion of the periphery of the carapace, are at first slightly bevelled off, and then vertically truncate; the sloping or bevelled part having the fine fibrous surface, which I have compared to coarse linen cloth. The vertical part of the border is slightly excavated in the fifth and sixth costal plates, but not so deeply as in the *Tri. circumsulcatus,* nor is the margin so thick in proportion to the length of the plate.

The neural plates are relatively smaller, in comparison to the costal plates, than in any of the foregoing species, but they agree in number; the eighth being suppressed, and the seventh reduced, in the same proportion as in *Tri. Henrici* and *Tri. Barbaræ,* by the median union of part of the seventh pair and of the eighth pair of costal plates. The fifth neural plate presents a simple oblong quadrilateral figure; the four neural plates in advance are six-sided, the two additional and shortest sides being formed by the truncation of the posterior angles; the sixth and seventh plates, on the contrary, have their anterior angles cut off. This modification in the form of the neural plates, and in their mode of juncture with the costal plates, relates to the opposite curvatures or inclinations of the costal plates, in the direction of the axis of the carapace: the anterior ones bending forwards, the posterior ones backwards, in addition to the curve common to all but the last pair of plates, transversely to the axis of the carapace, with the concavity downwards or towards the thoracic-abdominal chamber. The anterior internal angle of the second, third, and fourth costal plates is cut off; the posterior internal angle of the sixth, and both internal angles of the fifth pair of plates (*pl*5).

In these modifications of the form of the neural and costal plates, the *Tri. marginatus* agrees with the *Tri. Henrici* and *Tri. Barbaræ,* and differs from the *Tri. incrassatus,* in which all the neural plates but the seventh are six-sided, with the posterior internal angles truncated. Each of the costal plates, therefore, of the fifth pair, in the *Tri. incrassatus*, differs from those of the three other species of *Trionyx* here described, in having only the antero-internal angles truncated, instead of both these and the postero-internal ones.

In the form and general proportions of the first pair of costal plates, the *Tri. marginatus* shows an intermediate character between the *Tri. Henrici* and *Tri. Barbaræ ;* in the great breadth of the peripheral end of the seventh pair of costal plates it differs in a well-marked degree from both species, and especially from the *Tri. Henrici,* which it most resembles in its general contour. In the *Tri. incrassatus* the seventh costal plates maintain nearly an uniform breadth from end to end.

The antero-posterior diameter of each of the triangular plates of the last costal pair exceeds the transverse diameter, whilst these proportions are reversed in *Tri. Henrici ;* the difference in part depending on the different form of the posterior border of the carapace in the *Tri. marginatus*, which is truncated ; the free borders of the last costal plates forming a straight transverse line. The marginal pattern of the costal plates may be traced in a slighter degree round the neural plates.

The reticular sculpturing is better defined, and of a coarser pattern in the *Tri. marginatus* than in any of the previously defined species.

The middle line of the carapace is slightly depressed, as in the *Tri. incrassatus.* The general degree of convexity of the carapace, which is less than that in the *Tri. Henrici* and *Tri. Barbaræ*, agrees also with that of the *Tri. incrassatus.*

The length of the carapace from the fore part of the first neural plate is eleven inches ; its greatest breadth, across the suture between the third and fourth neural plate, is twelve inches.

This species is from the eocene deposit at Hordwell Cliff, Hampshire : it was discovered by the Marchioness of Hastings, and is preserved in her ladyship's collection at Efford House.

TRIONYX RIVOSUS. *Owen.* Tab. XVIII*A*.

This beautiful species of *Trionyx*, also discovered by the Marchioness of Hastings in the Eocene beds at Hordwell Cliff, has fortunately a characteristic pattern of sculpturing, which, like that in the *Tri. marginatus,* would serve for the determination of detached portions of the carapace. Any of the costal plates, for example, of the posterior half of the carapace, figured in Tab. XVIII*A*, might be distinguished by the sub-parallel longitudinal, and more or less wavy ridges, superadded to the more common reticulate sculpturing from the homologous parts of the carapace of any of the preceding fossil species of *Trionyx*, and, so far as I have yet seen, from any of the recent species.

The ridges in question, it will be understood, are longitudinal in respect of the entire carapace; they would be transverse to the long diameter of the detached costal plate; they become more wavy as they recede from the neural plates. Of these only the sixth ($s6$) has been preserved in the specimen described; it differs in shape from that in any of the foregoing species, in being broader in proportion to its length; its greatest breadth being, as in *Tri. Henrici*, *Tri. Barbaræ*, and *Tri. marginatus*, across its anterior fourth part. The fifth neural plate, as in the species above cited, has been an oblong quadrate one, the fourth plate has had its postero-internal angles cut off, contrariwise to the sixth. The fifth costal plates have accordingly the same character of truncation of both their internal angles, though less marked anteriorly. A portion of the seventh and the entire eighth neural plates have been superseded, as in the other fossil *Trionyces*, by the median growth and junction of the seventh and eighth pairs of costal plates.

In the forms and proportions of these plates the present species agrees best with the *Tri. Henrici* and *Tri. incrassatus;* the latter species differs from it by the breadth and convexity of the sixth costal plates (*pl6*). The smooth connate ribs (5, 6, 7), shown on the under surface of the costal plates, T. XVIII*A*, fig. 2, preserve a more uniform diameter, and do not expand in the degree shown in the *Tri. incrassatus*, T. XVII, 5, 6, 7; the rib (8) attached to the costal plate (*pl7*) is straighter in *Tri. rivosus* than in *Tri. incrassatus*.

The projecting extremities of the ribs are beautifully preserved in the specimen of *Tri. rivosus* here described: their greater length, as compared with those attached to the fifth, sixth, and seventh costal plates in *Tri. Barbaræ* (T. XVI*A*), depends upon the nonage of the present specimen, which is figured in T. XVIII*A* of the natural size.

The peripheral borders of the costal plates are bevelled off obliquely from above downwards, and project a little where they join the end of the subjacent rib; the surface of this is finely and longitudinally striated. The reticulate sculpturing of the carapace extends to the sloping peripheral border, as it does to the vertical thick border of the carapace in *Tri. incrassatus;* it is not separated from the border by a marginal decussating fibrous surface, as in *Tri. marginatus*, *Tri. Henrici*, and *Tri. Barbaræ*.

The longitudinal ridges of the carapace, which form the chief distinctive character of the *Tri. rivosus*, offer an interesting though slight approach to the main feature of the carapace of the Luth or coriaceous soft turtle (*Spargis coriacea*); but in this existing species the longitudinal ridges or carinæ are straighter and more elevated, and the surface of the carapace is smooth at the interspaces. The less parallel and wavy course of the ridges in the present extinct *Trionyx* give a sinuous course to the intercepted spaces, like the furrows left by streams of water which have temporarily coursed over a sandy surface, whence the name "rivosus" proposed for the species.

TRIONYX PLANUS. *Owen.* Tab. XIX *C*.

This species, like the *Tri. rivosus*, is represented by the posterior part only of the carapace, but the distinguishing characters are so well marked in it as to leave no doubt respecting the difference of the species from that of any of the above-defined *Trionyces*. The specimen consists of the last four pairs of costal plates, which are flat, with a coarse reticulate pattern on their upper surface, worn away towards the median end of the plates into a fossulate pattern, or detached pits; the reticulate sculpturing extends to the peripheral border of the costal plates, which is almost vertically cut down, and is scarcely at all produced where the attached rib projects: there is no marginal pattern along the anterior or posterior borders of the costal plate. The ribs are more neatly defined from the superincumbent costal plates than in any of the foregoing species, except, perhaps, the *Tri. rivosus*. The *Tri. planus* differs from them all in the complete obliteration of both the seventh and eighth neural plates, and by a partial obliteration of the sixth neural plate. This arises from a similar encroachment of ossification from the postero-internal borders of the sixth costal plates, upon the dermal cartilaginous matrix of the sixth neural plate, to that which happens in respect of the seventh neural and costal plates in the other *Trionyces*; whilst the whole of the seventh neural plate is superseded, as well as the eighth, and by the same encroachment of the corresponding pairs of costal plates.

These modifications and varieties of the osseous parts of the carapace are very significative of the essentially dermal nature of those parts, and show the small value and deceptive tendency of that developmental character on which Cuvier and Rathké have relied in pronouncing the neural plates to be developed spinous processes of vertebræ, and the costal plates to be expanded ribs. The connation of the seventh and eighth neural plates with the corresponding costal plates does not destroy their essential nature and existence, though it seems to make them part of the costal plates, any more than that connation with the neural arch in other *Chelonia* which seems to make them spinous processes.

Another distinctive character in the *Tri. planus*, as compared with the foregoing Eocene species, is the very close union, almost amounting to confluence, between the seventh and eighth costal plates of the same side, the original suture between which has been almost obliterated at their inferior surface.

In this character the *Tri. planus* resembles the *Tri. ferox*, Schweigger (*Gymnopus spiniferus*, Dum. and Bibr.), and *Tri. muticus*, Lesueur, but it differs from both by the flatness of its carapace, and the absence of any keel-like elevations upon its outer surface.

The middle of the posterior border of the carapace is slightly concave.

The specimen here described and figured was obtained by the Marchioness of

Hastings from the Eocene sand of Hordwell Cliff, and forms part of her ladyship's rich and instructive collection.

With the above portions of carapace, and apparently belonging to the same species of *Trionyx*, were found the two osseous plates, naturally and suturally united together, which are figured in T. XIX*D*, fig. 6, *hs, ps* ; they present a similar coarse reticulate pattern on their external surface, with the same tendency to a concentric arrangement of the raised parts towards the periphery of the plate ; their inner surface is smooth, slightly undulating, but upon the whole a little concave, and without any indication of adherent ribs. I regard them therefore as parts of the plastron, and they agree best with the hyosternal and hyposternal elements of the right side ; yet differ in having no tooth-like processes extending from the inner border, which is convex instead of being concave, where the two elements join each other.

At the inner and anterior angle of the hyosternal there is, however, the fractured base of what was probably a tooth-like process ; and there is similar evidence of such processes having extended from the posterior angle of the hyposternal, close to what I take to have been part of the notch for the xiphisternal.

These fragments at least show that the *Tri. planus*, or whatever species from Hordwell they belonged to, must have had a very different form of plastron from that of the *Tri. incrassatus* of the Isle of Wight, of which the conjoined hyosternal and hyposternal bones are figured in T. XVII*A*, figs. 3 and 3′, and from that plastron of which the hyposternal piece, from Bracklesham, is figured in T. XIX*D*, fig. 7.

TRIONYX CIRCUMSULCATUS. *Owen.* Tab. XIX*B*, figs. 1, 2, and 3.

It may seem to have been hazarding too much to found a species on a single character when manifested by a single fragment of a carapace, which is all that at present represents such species ; yet the character in question is so strongly marked, and so different from that of the same part of the carapace of any other fossil or recent species of *Trionyx*, that there appears to be no other alternative than to regard it as specific. The character in question is the groove or canal which is excavated in the thick vertical margin of the expanded free extremity of the fourth costal plate of the left side, figured in T. XIX*B*, figs. 1, 2, and 3. The vermicular sculpturing of the external surface of this plate, and its proportions and connexions with the connate rib, prove it to belong to the carapace of a *Trionyx*.

Previously to receiving this specimen from Lady Hastings, my attention had been drawn to the different modes in which the extremities of the costal plates of the different species of *Trionyx* were modified, in order to form the border of the carapace ; sometimes obliquely bevelled down to an edge, as in the *Tri. Barbaræ* and the fragment of the *Trionyx pustulatus*, from Sheppy, figured in T. XIX*B*, 7—10 ; sometimes cut down vertically, or nearly so, as in the thickened border of *Tri. incrassatus;* sometimes with a marginal modification of the external sculpturing before the edge was formed, as in *Tri. marginatus;* sometimes without any such border-pattern, as in *Tri. rivosus.* But whatever character the border of a carapace has presented, has been constant in the same species, in which it is modified only at the fore part of the border formed by the nuchal plate, and at the back part formed by the short and small eighth pair of costal plates.

From this, therefore, it is to be inferred that the peculiar modification presented by the free border of the fourth costal plate, T. XIX*B*, fig. 3, was repeated in all the other costal plates, excepting, perhaps, the last pair; and consequently that the carapace was almost entirely surrounded by a thick, vertical border, deeply grooved,—a character which is expressed by the specific name *circumsulcatus,* selected to denote the Eocene *Trionyx,* represented by the fragment of the carapace here described.

This fragment, which consists as before said of the fourth costal plate of the left side, presents the common reticulate pattern of its external sculptured surface, but with some modifications not presented by the before-described species; the meshes are smaller near the ends of the plate than at its middle part, and the network is finest near the peripheral end. In the *Tri. marginatus* more particularly, and in a minor degree in *Tri. incrassatus,* the *Tri. Henrici,* and *Tri. Barbaræ,* we observe the raised parts of the network assuming a linear arrangement, more or less concentric, with the circumference of the carapace; but there is nothing of the kind observable in the *Tri. circumsulcatus.* In this species also the outer surface of the costal plate presents a distinct though slight double curvature; the usual convexity being changed into a concavity near the peripheral border: and, as the inner surface presents the usual uniform concavity, the peripheral part of the plate suddenly augments in thickness as it approaches the grooved border. (See fig. 2.) The character which distinguishes the *Tri. incrassatus* from *Tri. Henrici, Tri. Barbaræ,* and *Tri. rivosus,* is exaggerated in *Tri. circumsulcatus,* and there is added to it the groove, of which there is no trace in *Tri. incrassatus,* and but a feeble one in the fifth and sixth plates of *Tri. marginatus.*

The connate rib is almost wholly sunk into the substance of the superincumbent costal plate in the *Tri. circumsulcatus;* it is less prominent than in any of the foregoing species, especially at its distal part, which is also less expanded than in the *Tri. incrassatus.* The free extremity of the rib is entire, and is very short, as is shown in figure 1.

TRIONYX PUSTULATUS. Tab. XIX*B*, figs. 7, 8, 9.

The contrast which the fragment above referred to, of apparently the homologous costal plate to the one last described, presents in the character of its peripheral border, and in the prominence of the connate extremity of the rib on its under surface, is so great, as must impress the value of such characters upon the palæontologist. The outer surface of the present fragment presents a well-marked reticulate, or rather pustular, pattern, but a coarser one than in the *Tri. circumsulcatus.* The reticulation is continued to the beginning of the bevelled border in fig. 7, which slopes gradually to an edge; beneath which the free end of the rib projects. The *Tri. rivosus* most resembles the present fragment in this character.

The fragment is from Sheppy. I strongly suspect it to belong to a species distinct from any of those from Hordwell; and, in the hope of acquiring more illustrative specimens, the attention of collectors is directed to it by the specific name and the figure here given.

TRIONYX. *Sp. ind. Bracklesham.*

The left hyposternal bone of the *Trionyx* from Bracklesham (figured in Tab. XIX*D*, fig. 7) resembles that from the Hordwell Eocene, referred to *Trionyx planus* (fig. 6, *ps*), in the convexity of the inner border at that part where it is concave in the *Tri. incrassatus* (T. XIX, fig. 3, *ps*); but it differs from the *Tri. planus* in being uniformly convex as far as the xiphisternal notch, and is not indented before forming that notch, as it is in the *Tri. planus* (T. XIX*D*). The present hyposternal shows also very plainly the base of a fractured tooth-like process of the subjacent hæmapophysis projecting from the inner border, where there is no such trace of a process in the *Tri. planus*. There are also the bases of a tooth-like process on both sides of the xiphisternal notch, and at the posterior outer angle of the hyposternal bone. The external border of the bone in advance of these processes is longer and straighter than in the corresponding part of the hyposternal of the *Tri. incrassatus*.

The species of *Trionyx* from Bracklesham cannot, however, be safely defined until the characters of its carapace are known. The present specimen forms part of the valuable and instructive collection of Frederick Dixon, Esq., F.G.S.

Family—PALUDINOSA.

This family, if regard were had to the number of species it contains, might be deemed the typical one of the order *Chelonia.* But in the series of extinct species, from the particular formation of Great Britain, to which the present Monograph is restricted, the number of marsh tortoises is small in comparison with those that were more truly aquatic (*Fluvialia*), and which inhabited the sea (*Marina*); and such a result might have been anticipated from the nature of their matrix, as it is elucidated by other classes of fossil animals, the remains of which are found in the London clay.

The feet of the *Paludinosa* have the digits comparatively free; more than three toes, as in *Tetronyx*, Lesson, and usually all five, are armed with claws, and are united together by a web only at their base; but the extent of this web and the length and flexibility of the digits vary in the different species and sub-genera, and accordingly they manifest various degrees of aptitude for swimming, or for climbing the banks of the streams or marshes which they habitually frequent, and for walking on dry land.

The costal plates extend, in the mature individuals, to the ends of the ribs, and articulate with the marginal plates; the dermal pieces of the plastron are co-extensive with the abdominal integument, and unite together by suture so as to form an unbroken expanse of bone; the sides of which, formed by part of the hyosternals and hyposternals, unite with a corresponding proportion of the lateral borders of the carapace. There is a gradation in the degree of convexity of the carapace, and in the angle at which the sides of the plastron bend up to join the carapace, which progressively brings the marsh tortoises nearer to the true land tortoises (*Terrestria*), and some of the steps in this progression of affinities are illustrated by the fossils from the London clay.

Those that, by the flatness of their carapace and plastron, depart least from the fluviatile forms of the order will be first described.

Genus—PLATEMYS.

PLATEMYS BULLOCKII. *Owen*. Tab. XXI.
Report on British Fossil Reptiles, Trans. British Association, 1841, p. 164.

Amongst the fossil Chelonians of the London clay, the portable dwelling-house of which was provided with side walls as well as a floor and roof, are some tolerably large species, remarkable for the lowness of the roof of their abode, and especially for the flatness of its floor.

A rigid comparison of the numerous species of the marsh-dwelling Chelonians, which the active researches of naturalists have brought within the domain of science, has led to their classification into several groups, to which generic or sub-generic names are attached, and the fine preservation of the characteristic part of the skeleton of the specimen from Sheppy, figured in Tab. XXI, gives the opportunity for determining to which of these subdivisions of the genus *Emys* of Bronguiart that specimen belongs.

In my 'Report on British Fossil Reptiles,' the result of these comparisons, as regards the present fossil, were simply indicated by the sub-generic name, and I confined myself to a description of the specific distinctions noticeable in the only example I had then seen.

The present species differs from all those to which MM. Dumeril and Bibron restrict the term *Emys*,* by the presence of a thirteenth scute—the intergular one (*ig.* T. XXI) upon the plastron ; from the genera *Cistudo* and *Kinosternon* it differs by the absence of any moveable joint between the parts of the plastron ; from the *Tetronyx* by the rounded anterior border of the plastron, and the greater number of scutes that have left their impressions upon it : it resembles the genus *Platysternon* in the flatness of the plastron and the horizontality of its lateral prolongations ; but it differs from the only known species of that genus in the contour of the sternum, which is elliptical and rounded in front, and has the lateral prolongations one third the length of the entire sternum. It has also the intergular scute, which is absent in the *Platysternon*, as in the *Emydes* of Dumeril and Bibron. The presence of this scute, so plainly indicated at *iy* in the petrified plastron from Sheppy, together with the impressions of six pairs of the more constant scutes of the plastron, indicate that the depressed form of the probably estuary terrapene to which that plastron belonged, has appertained to the section which the eminent French Erpetologists above cited have called *Pleurodères*, or those that could retract their neck beneath the side only of the anterior aperture of their thoracic abdominal case.

From the genus *Peltocephalus* the fossil under comparison differs by the marginal position of both gular (*gu*) and intergular (*ig*) scutes, and by the slight narrow emargination of its posterior extremity (*xs*). An outline of the natural size of this emargination is added in the plate.

It more nearly resembles the *Podocnemys expansa* in the forms and proportions of the plastron scutes ; but the three anterior ones (*gu, gu,* and *ig*), are not wedged in (*enclavées*) between the humeral scutes (*hu*), but are on a plane anterior to them.

The form and proportions of the plastron in certain species of the *Platemys*, Dumeril and Bibron, and the number and relative position of the scutes which covered it, offer the nearest resemblance to those of the present fossil, and, with the results of the foregoing comparisons, have determined my reference of the specimen in question to that genus.

Like the *Platemys Spixii* (*Emys depressa* of Spix), *Platemys radiolata, Platemys gibba*, and some others of the genus, the sternum is rounded at its anterior border, and notched at its posterior and narrower extremity.

The intergular scute (*ig*) which crosses the median suture of the episternals (*es*) is sub-pentangular and larger than either of the gular pair ; its point encroaches a little upon the entosternal bone (*s*). The gular scutes (*gu*) are triangular, and, with the intergular one, cover the anterior border of the plastron.

The humeral or brachial scutes (*hu*) are inequilateral quadrate plates ; the pectoral scutes (*pe*) and the abdominal scutes (*ab*) are transversely oblong and quadrate. The femoral scutes are inequilaterally quadrate, the posterior external angles being prolonged and rounded off. The anal scutes would be sub-rhomboidal were the posterior

* Erpétologie Générale, 8vo, 1835, tom. ii, p. 232.

end of the plastron entire. There are impressions of three scutes—the axillary, the inguinal, and a supplementary one,—upon each lateral prolongation of the plastron, covering the suture between this and the marginal plates of the carapace (*aa*), in which the present fossil resembles the *Platysternon* or large-headed *Emys* of China; but the lateral walls are relatively longer, being equal in antero-posterior extent to one third the same diameter of the entire plastron; whilst in the subgenus *Platysternon* they are less than one fourth. The general form of the plastron is also very different; in the *Platysternon megacephalum*, e. g. the plastron has an oblong quadrilateral figure, with an open-angled notch behind.

Retaining, then, the present species in the genus *Platemys*, as defined by Duméril and Bibron, we find that it enters into that small minority of the group in which the plastron is rounded instead of being truncate anteriorly.

In the present remarkable fossil the plastron forms almost a long ellipse, the hinder, division being very little narrower, but tending to an apex, which is cut off by a shallow emargination. The lateral walls, of the length above defined, extend outwards almost parallel with the plane of the sternum, and expand to join by a wavy or rather zigzag suture the marginal plates; six of these (*a a a a a a*) are preserved on each side; their lower sides form a very open angle with the lateral walls: but the fractures of these parts indicate that their horizontality may be in part due to accidental pressure.

The anterior part of the entosternal (*s*) is bounded by two nearly straight lines, converging forwards at an angle of 65°, with the apex rounded off; the posterior contour of this bone is nearly semicircular. The length of the entosternal is two inches ten lines; its breadth three inches seven lines; the forms and relative positions of the other elements of the plastron are sufficiently illustrated by Tab. XXI: *es, es* marks the extent of the left episternal; *hs, hs* are the hyosternals; *ps* the hyposternals; *xs* the xiphisternals.

The chief peculiarity of this plastron is the intercalation of a supernumerary piece of bone, bearing the letters *pe* and *ab* between the hyosternal and hyposternal elements on each side; so that the middle third of the plastron is crossed by two transverse sutures instead of one; each suture being similarly interrupted in the middle by an angular deflection from the right, half an inch back, to the left side.

The extremities of the transverse sutures terminate each at the apex formed by the inner or lower border of the parallel marginal plates. The first or anterior of these sutures is distant from the anterior margin of the plastron six inches five lines; the second suture is distant from the same margin eight inches nine lines; the right half of the suture, which is a few lines in advance of the left, is the part from which these measurements are taken.

Since this deviation is rare, it having been noticed for the first time in the original description of the present specimen, a naturalist, not having the specimen at hand for comparison, might at first be led to suspect that the transverse impressions of the second (pectoral) or third (abdominal) pairs of scutes had here been mistaken for a suture; but due care was observed to avoid this error; the scutes of the plastron have left obvious impressions at *pe, fe,* which prove that they were in the same number as in the Platemydians generally, and were quite distinct from the sutures in question.

Thus the intergular scute (*ig*) is in the form of an ancient shield; the gular scutes (*gu*) are small inequilateral triangles, with their posterior border parallel with that of the succeeding pair of scutes. The posterior transverse boundary of these,—the humeral scutes (*hu*)—crosses the plastron four inches and a half from its anterior margin; that of the pectoral pair of scutes crosses at seven inches and a half from the anterior border, and between the two transverse sutures; that of the abdominal pair (*ab*) at ten inches distant from the anterior margin, and about one inch and a quarter behind the second transverse suture; passing straight across the plastron between the posterior concave margins of the lateral wall. The posterior boundary of the fifth or femoral pair of scutes (*fe*) inclines obliquely backwards from the median line, as usual; it is three inches behind the preceding transverse impression.

It is in the interspace of these impressions that traces of the transverse suture between the hyposternals and xiphisternals are obvious, about four inches from the posterior extremity of the plastron. If these traces were not so obvious, it might be supposed that the xiphisternals were of unusual length, entering into the formation of the lateral wall, and extending backwards from the second transverse suture to the end of the plastron; but this disproportion would be hardly less anomalous than the existence of the additional pair of bones intercalated between the hyo- and hyposternals which the present fossil evidently displays.

In most of the existing large *Emydes* and *Platemydes*, the median transverse suture traverses the plastron a little behind the third pair of scutes, and so crosses the fourth or abdominal pair (*ab, ab*); and according to this analogy, the second transverse suture in the fossil agrees with the single one ordinarily present, and has most right to be regarded as the normal boundary between the hyo- and hyposternals. One of the most distinctive characters of the present extinct *Platemys* is, therefore, the division of each hyosternal into two, the plastron consisting of eleven instead of nine pieces; if the very interesting anomaly which it displays be not an accidental or individual variety. Viewed in the latter light, its explanation is suggested by that homology of the hyosternals and hyposternals which determines them to be connate and expanded abdominal ribs (hæmapophyses), and thus we may view the oldest of the known Platemydians as exhibiting, like many other extinct forms, a nearer approach to the more typical condition of the abdominal ribs, as they are shown, e. g. in the *Plesiosaurus.* Whereas, on Geoffroy's hypothesis, that the plastron is the homologue of the sternum of the bird, it would be a further deviation from that type.

9

The fine example of *Platemys Bullockii*, here described and figured, was purchased for the British Museum at the sale of Mr. Bullock's collection.

I am happy in the opportunity of expressing my acknowledgments to Charles König, K.H., F.R.S., for the urbanity with which every requisite facility was afforded.

PLATEMYS BOWERBANKII. *Owen.* Tab. XXIII.

Report on British Fossil Reptiles, Trans. British Association, 1841, p. 163.

This species is represented by a fine specimen exhibiting not only the plastron (fig. 2), but likewise a great portion of the carapace (fig. 1), from Sheppy, in the rich collection of the fossil remains from that island in the possession of J. S. Bowerbank, Esq., F.R.S. It equals in size the *Platemys Bullockii*, in the British Museum, but differs in the absence of the finely punctate character of the exterior surface of the bones; in the greater antero-posterior extent of the lateral walls, and the longer curves which they form in extending from the body of the plastron.

The carapace (fig. 1) presents the same equality of breadth of the neural plates ($s2$—$s7$) as in the *Emys testudiniformis;* but they diminish more rapidly in length as they recede in position; and the whole carapace is much more depressed; it is flat along its middle tract. The sixth neural plate ($s6$) is a hexagon of nearly equal sides; the seventh ($s7$) is a pentagon; the mesial or vertebral ends of the seventh pair of costal plates ($pl7$) meet and unite behind it, so as to conceal or supersede the eighth neural plate. In the circumstance of the neural plates decreasing in length without losing breadth, as well as in the mutual junction of the seventh costal plates, the present fossil resembles the Sheppy carapace from Mr. Crow's collection, which Cuvier has figured, and which may, therefore, have belonged to the present species of *Platemys*.

The plastron (fig. 2) is thirteen inches in length and ten inches in breadth; it is rather broader before than behind, rounded at the anterior border, with a shallow emargination at the middle of the posterior border, but wider than in the *Platemys Bullockii*, and with the angles on each side rounded off. The under surface is nearly flat, slightly convex at the fore part, and as slightly concave behind. The lateral walls uniting the plastron to the carapace are five inches in antero-posterior extent.

The entosternal (s) resembles that of the *Platemys Bullockii* in general form, but is longer than it is broad, instead of the reverse proportions. The two anterior sides meet at a right angle. The episternals (es) are broadest behind. The middle part of the plastron is almost equally divided between the hyosternals (hs) and hyposternals (ps). There is a trace of the intercalary piece (hp), which is seen extending across the plastron of the *Platemys Bullockii;* here it is wedged into the outer interspace of those bones, like one of the external portions of the composite abdominal ribs in the Plesiosaur. In the relative length of the lateral walls the *Platemys depressa* most resembles the present species.

Genus—EMYS.

EMYS TESTUDINIFORMIS. *Owen.* Tab. XXIV.
Report on British Fossil Reptiles, Trans. British Association, 1841, p. 161.
EMYS DE SHEPPY. *Cuv.* (?)

From the preceding genus of the *Chelonia paludinosa* the present species differs in the depth of the bony cuirass, the convexity of the carapace, and the concavity of the plastron (T. XXIV, fig. 6). The more immediate affinities of the present fossil are elucidated by the comparison of the points of structure which it displays with the anatomical characters of the carapace of the *Platemys* and *Testudo*.

The specimen, on which the species here called *Emys testudiniformis* is founded, includes a large proportion of the first, second, third, fourth, fifth, and sixth, with a fragment of the seventh costal plates of the left side; a small proportion of the second, third, fourth, fifth, and sixth neural plates; the hyosternals and hyposternals, and part of the entosternal bones of the plastron.

The first costal plate is one inch ten lines in greatest breadth, one inch five lines broad at its junction with the neural plates, and four fifths of the vertebral margin is articulated with the second neural plate; one fifth part, divided by an angle from the preceding, joins a corresponding side of the lateral angle of the third neural plate; in this structure it resembles both the genus *Testudo* and some species of *Emys*.

The third, fourth, fifth, and sixth neural plates are of equal broadth, as in *Emydes*; not alternately broad and narrow as in the *Testudines*; they are likewise of uniform figure, as in most *Emydes*; not variable, as in *Testudines*; the neural plates also resemble those of the existing *Emydes*, and particularly of the Box-terrapin (*Cistudo*) in form. The lateral margin of each is bounded by two lines, meeting at an open angle, the anterior line is only one fourth part the length of the posterior one; and this resemblance may be stated with confidence, since the portion of the entosternal piece preserved in the plastron determines the anterior part of the fossil.

The costal plates preserved in the present Chelonite differ from the corresponding ones of the tortoises, and resemble those of the *Emydes* in their regular breadth, and the uniform figure of the extremities articulated with the vertebral pieces; the anterior line of the angular extremity is nearly three times as long as the posterior one.

Further evidence of the relation of the present Chelonite to the fresh-water family is given by the impressions of the epidermal scutes; those covering the vertebral plates (*scuta vertebralia*) agree with those of most *Emydians* in the very slight production of the angle at the middle of their lateral margins, which is bounded by a line running parallel with the axis of the carapace, except where it bends out to form that small angle.

The middle part of each side of the plastron, in the *Emys testudiniformis*, is joined to the carapace by a strong and uninterrupted bony wall, continued from a large proportion of the hyosternal and hyposternal bones upwards to the marginal costal pieces. The median margin of the hyosternals and hyposternals are articulated together by a linear suture, traversing the median line of the plastron, and only broken by a slight angle formed by the right hyposternal, which is a little larger than the left. A similar inequality is not unusual in both tortoises (*Testudinidæ*) and terrapenes (*Emydidæ*). The transverse suture is, of course, broken by the same inequality; that portion which runs between the left hyosternals and hyposternals being two or three lines in advance of the one between the right hyosternals and hyposternals. The posterior half of the broad entosternal piece is articulated to a semicircular emargination at the middle of the hyosternals; so that the whole plastron forms one continuous plate of bone. This is relatively thicker than in existing *Emydes,* resembling in its strength that of tortoises; and it is likewise slightly concave in the middle, which structure is more common in tortoises than in Emydians, save those in which the sternum is moveable; in most of the other species the sternum is flat or slightly convex.

I have shown in my paper on the Turtles of Sheppy,* that the carapace figured by Cuvier† was not sufficiently perfect to decide the affinities of the Chelonian to which it belonged; if the vertebral scutes were less broad and angular than in marine turtles, the neural plates—much less variable in their proportions—were, on the other hand, as narrow as in turtles. But with reference to the plastron of the Sheppy Chelonite, figured by Parkinson,‡ and supposed by Cuvier to belong to an *Emys* of the same species as the carapace above alluded to, I have been able to determine, by an examination of the original specimen in the museum of Professor Bell, that it belonged to the marine genus *Chelone* and to the species *longiceps*. In the fossil *Emys* in Mr. Bowerbank's collection, the plastron being in great part preserved, establishes its nonconformity with the marine turtles, and manifests a striking difference from Parkinson's fossil plastron.

The entosternal piece is impressed, as in Tortoises and Emydes, by the median longitudinal furrow, dividing the two humeral scutes; the transverse linear impression dividing the humeral from the pectoral scutes traverses the hyosternals half an inch behind the suture of the entosternal; the second transverse line, which divides the pectoral from the abdominal scutes, is not so near the first as in tortoises, but bears the same relation to the transverse suture of the plastron as in most Emydes; it does not pass straight across the plastron, but the right half inclines obliquely inward to a more posterior part of the median suture than is touched by the left half. The third transverse line, which divides the abdominal from the femoral scutes, passes straight

* Geological Proceedings, December 1, 1841.
† Ossemens Fossiles, tom. v, part iv, pl. 15, fig. 12.
‡ Organic Remains, vol. iii, pl. 18, fig. 2.

across the plastron between the posterior ends of the bony lateral walls, uniting the carapace and plastron.

	Inches.	Lines.
The breadth of the plastron is	5	10
The outer posterior extent of the lateral wall is . . .	3	9
The breadth of the entosternum	1	5
The depth of the whole bony cuirass at the middle line is .	4	0

In the convexity of the carapace and relative depth of the osseous box, the Sheppy Chelonite slightly surpasses most existing species, resembling in this respect the *Emys ocellata* and *Cistudo Carolina.* The plastron is also slightly concave, as in the male of *Cistudo vulgaris:* it is, however, entire at the line where the transverse joint of the plastron exists in the box-tortoises; and the extent and firm ossification of the lateral supporting walls of the carapace forbid likewise a reference of the fossil to those genera.

The general characters of the present fossil, more especially the uniformity of size and breadth of the preserved vertebral plates and ribs, prove it to be essentially related to the fresh-water or Emydian Tortoises. It exceeded in size, however, almost all known Emydians, and was almost double the dimensions of the Emydian species (*Cistudo Europea*) now inhabiting central Europe. It appears, like the *Cistudines,* to have approached the form of the land tortoises, in the convexity of the carapace, but without possessing that division and hinge of the plastron which peculiarly distinguishes the box-tortoises. The contraction of the anterior aperture of the bony cuirass, especially transversely as compared with the *Platemydians,* would indicate more restricted powers of swimming, and consequently more terrestrial habits. In the thickness and strength of the bones of the buckler, especially of the sternum, we may discern an approach to the genus *Testudo.*

Assuming that the Chelonite here described may be identical with that of which the carapace from Mr. Crow's collection is figured in the 'Ossemens Fossiles,'[*] the " *Emys de Sheppy*" of Cuvier will be one of the "synonyms" of the present species. Mr. Gray, in his 'Synopsis Reptilium,' 8vo, 1831, has given Latin names to all the fossil reptiles indicated or established by Cuvier, and has called the " *Emys de Sheppy*" " *Emys Parkinsonii,*" referring as representations of this species, not to the figure of the carapace above cited, which may belong to the same species as the present *Emys,* but to the figure of the plastron, copied by Cuvier from Parkinson's 'Organic Remains,' and to the figure of the skull in the same work, both of which most unquestionably belong to the genus *Chelone* and not to the genus *Emys.*

The " *Emys Parkinsonii*" of Mr. Gray is a synonym of my *Chelone longiceps.* Cuvier's name,—which, besides the claim of priority, is the result of laborious and direct comparison devoted to the elucidation of its subject,—if rendered into Latin would be *Emys toliapicus;* but as the species to which it refers may not be the one

[*] Ed. 1824, vol. v, part ii, pl. 15, fig. 12.

here described, and is by no means the only fresh-water tortoise which the clay of Sheppy has yielded; and since the characters of the present species have not hitherto been defined nor its affinities to the land tortoises been pointed out, the interests of science appeared to me to be best consulted by giving a distinct name to the present species.

The fossil here described is from the Eocene clay of Sheppy Island, and forms part of the collection of J. S. Bowerbank, Esq., F.R.S.

EMYS LÆVIS. *Bell.* Tab. XXII.

The only specimen I have seen of this species, I obtained from Sheppy a few months since, and it is now in my collection. It has some remarkable peculiarities which distinguish it, at first sight, from every other species of Emydian, either recent or fossil.

The specimen is imperfect at each extremity; the carapace wanting anteriorly the nuchal plate, and posteriorly from the eighth neural plate inclusive. The contour of the carapace is remarkably even, free from all inequalities of surface, and forming, from side to side, nearly a perfect segment of a circle, uninterrupted by either carina or depression of any kind. The whole surface of the bone also is remarkably smooth.

The first neural plate (fig. 1, $s1$) is narrow, being not more than two fifths as broad as it is long; the sides parallel for the first two thirds of its length, then slightly narrowed; its sides are not interrupted by the costal sutures, as the posterior margin of the first costal plate ($pl1$) joins the anterior part of the second neural. The second, third, and fourth neural plates ($s2$—$s4$) are of an elongated hexagonal form, and nearly resemble each other; the fifth, sixth, and seventh ($s5$—$s7$) are also hexagonal, but each shorter than the preceding one; the sixth is narrowed somewhat abruptly, and the seventh still more so, the latter being also shorter than it is broad.

Although the posterior part of the carapace is considerably broken, there appears evidently to be an interval between the seventh and eighth neural plates; at which part the posterior portion of the seventh costal plate and the anterior portion of the eighth approximate to the corresponding plates of the opposite side, on the median line, without the intervention of the neural plates; a peculiarity which I do not remember to have seen in any other of the *Emydidæ*.

The first costal plate occupies in its breadth the whole length of the first neural, and the anterior fifth only of the second; but in consequence of the gradual shortening of the neural plates in the portion of each, posterior to the angle at which the costal sutures join them, the seventh neural receives the costal suture at about the middle of its length.

The marginal plates (fig. 3, *a*, *a*, *a*) are broad, smooth, and curved evenly to the edge, where they turn under at nearly a right angle.

The second and third vertebral scutes(*v*2, 3) are twice as broad as they are long, the outer angles being nearly right angles; and this must be, to a great extent, a permanent character, as the specimen is evidently not young. The fourth vertebral scute (*v*4) is hexagonal, and its breadth is about one fourth greater than its length.

Of the plastron (fig. 2), the whole of the anterior portion is wanting, including the entosternal, the episternals, and a portion of the hyosternals; and the posterior portion has lost the greater part of the xiphisternals. The bones which remain form a broad, somewhat convex, uniform surface.

The most remarkable circumstance connected with this part of the osseous box is the existence of a pair of intercalated, irregularly-formed bones (*hp*), which stand between the marginal portion of the hyosternal (*hs*) and hyposternal bones. These would appear to represent the pair of additional bones which will be seen in *Platemys Bullockii* (Tab. XXI), stretching across between the hyosternals and hyposternals, and, in the latter case, meeting like them in the median line.

I have examined many skeletons of Emydes, but have never observed any similar structure in this genus; but in the genus *Terrapene,* including the ordinary box-tortoises, there appears to be, in some cases, a rudiment of a corresponding bone.*

The total length of the carapace of this specimen, judging from comparison with perfect recent examples of the same genus, was probably rather more than eight inches, and its breadth is six inches.

T. B.

EMYS COMPTONI. *Bell.* Tab. XX.

The beautiful specimen of fresh-water Chelonia which forms the subject of the present description, is in the collection of the Marquis of Northampton, who has kindly allowed me the use of it, and to whose respected name I have dedicated it.

The general form of this species, as well as many details of its structure, is so similar to that of a typical land tortoise, that it is difficult at first to reconcile its aspect with the idea of its being at all aquatic in its affinities. It is, however, doubtless a true Emys; and although the present specimen is a young one, its characters are sufficiently marked to enable us to distinguish it from every other. The costal plates

---

* The sternal bones appear liable to occasional curious anomalous variations. Thus, while in *Platemys* there is a perfect pair of intercalated bones between the hyosternals and the hyposternals, and in the present species an approach to a similar interpolation, we find, on the contrary, in *Gymnopus,* a genus of Trionychidæ, the only skeleton of which in this country I have now in my possession, the hyosternals and hyposternals constitute but a single bone on each side, a peculiarity which I believe to be perfectly unique in the whole of the Chelonian order. [T. B.]

had not become ossified to the extremity of the ribs, and there is consequently a space between the costal and marginal plates, interrupted by the free extremities of the ribs, which just reach to the marginal plates. It is the only specimen of the family which I have seen, amongst the fossil Chelonian remains, in which the whole series of neural plates, with the nuchal and pygal, remain without material injury; and the plastron is also nearly entire.

The nuchal plate (fig. 1, *ch*) would form a triangle with its posterior angle obtuse, but that this angle is truncated for its articulation with the first neural (s_1). This latter plate is quadrate, a little longer than broad, and rather narrowed forwards. The second (s_2) and third (s_3) are also quadrate, and nearly equilateral. The fourth (s_4) is, however, rendered hexagonal by the termination of the costal suture at a short distance from the anterior margin; it is quite as broad as it is long. The fifth neural plate (s_5) is of a similar form, but notably longer than it is broad, forming a broad hexagon, with the lateral angles nearer the anterior than the posterior margins. The seventh (s_7) is the only one which forms a nearly symmetrical hexagon, broader than it is long, but with the lateral angles equidistant from the anterior to the posterior margins. The eighth and ninth neural plates (s_8, 9) are regularly quadrate, the former being broader than it is long, the latter forming a perfect square. It is very remarkable how much more closely the seventh and following neural plates to the tenth are united than any of the anterior ones; indeed the sutures between the seventh and eighth, and between the eighth and ninth, are with difficulty observable, notwithstanding the youth of the individual. The tenth neural (s_{10}) and the pygal (p) plates are somewhat injured and bent down abruptly by some violence.

I have dwelt somewhat in detail upon the direction of these plates, as their characters evidently bear upon the near relation of this species to the terrestrial type already alluded to.

The internal margin of the first costal plate (pl_1) exactly coincides with the length of the first neural. The second and fourth costal plates (pl_2, 4) expand towards the margin of the carapace, and the third and fifth (pl_3, 5) become narrower in the same direction in a similar degree.

The marginal plates present no important peculiarity in this young specimen.

With regard to the impressions left by the horny scutes, we find that although they are of the ordinary general form, they are less broad and spreading in proportion to their length, than is ordinarily the case in the *Emydidæ*, and particularly in immature age; thus offering another character approaching the terrestrial type.

The plastron (fig. 2) is tolerably perfect, and presents the remarkable expanse which ordinarily characterises the land and fresh-water forms, but especially the former; and the anterior and posterior openings between the carapace and the plastron, for the exit and play of the extremities, are somewhat contracted, and thus appear scarcely to afford sufficient room for the natatorial habits of an aquatic species.

The entosternal plate (*s*) forms an almost regular rhomb; the episternals (*es*) are much broken, and offer no peculiarity in the parts which remain; nor is there, in the general form of the hyosternals (*hs*) or hyposternals (*ps*), or the xiphisternals (*xs*), anything which calls for particular notice.

The contour of the bony case, viewed as a whole, bears out the close relation to the terrestrial form which I have assigned to this species. The slightly curved costal regions of the carapace, and the even flatness of the vertebral portion, as well as the outline of the dorsum, when viewed laterally, show a very striking approximation to the small African species of true Testudo, *T. areolata*, and still more to *T. signata*. But if its geological position did not of itself preclude our considering it as belonging to a terrestrial group, the structure of many parts of its osteology would be sufficient to justify our considering it as a true Emydian.

	Inches.
Length of the carapace	3·2
Breadth of ditto	2·9
Height of the bony case	1·3

T. B.

EMYS BICARINATA. *Bell.* Tab. XXV and XXVI.

The specimen before me, the only one which I have yet met with of this species, is very large, and, from the close union of the bones, and the nearly obliterated condition of the sutures, is evidently of considerable age; a fact also attested by the forms of the vertebral scutes (*v2, v3, v4*, T. XXV), which have become greatly narrowed in proportion to their length.

The general outline of the carapace must have been nearly orbicular. The elevation moderate; the part occupied by the vertebral scutes, and about half an inch on each side of them, flattened; and this plain portion bounded on each side by a low obtuse carina, which is itself obscurely and irregularly grooved longitudinally. The sides are considerably sloping, with but a slight curvature.

The carapace is wanting anteriorly in nearly the whole of the nuchal plate, and posteriorly from the tenth neural inclusive. At the sides a few fragments only of the marginal plates exist.

The first neural plate (*s1*) is nearly oval, and, as usual in this family, is wholly included within the first pair of costal plates; it is considerably longer than any of the succeeding ones. The second neural (*s2*) is nearly as broad as it is long, the anterior angles truncated as usual, posteriorly somewhat narrowed; the third neural (*s3*) has the peculiarity of being longer than even the second, and is less narrowed behind; the fourth to the seventh inclusive (*s4-7*) are gradually shorter, the seventh forming a broad hexagon, with the lateral angles (meeting the costal suture) nearly midway between the anterior and posterior margins. The eighth (*s8*) is also broader than it is

10

long, but the lateral angle is near the anterior margin, as in the preceding plates. The ninth (*s9*) is somewhat expanded posteriorly, but less so than usual.

The sixth and seventh of the neural plates are considerably raised towards the centre, but with a slight longitudinal depression along the median line; and there is a considerable triangular or wedge-shaped elevation, commencing with its base near the anterior margin of the eighth, and extending to the posterior margin of the ninth neural plate.

The costal plates (*pl*1—8) differ from those of the species in general in being more regularly parallel at their lateral margins.

The first vertebral scute reaches to the posterior third of the first neural plate (*v1*), and its lateral margins are expanded forwards, but with a slight curve. The second and third (*v2, 3*) have nearly parallel sides, and are both longer than they are broad, the lateral angles being extremely inconsiderable; the fourth (*v4*) is hexagonal, but still with short lateral angles; the fifth (*v5*) has the lateral margins, and, as usual, becomes broader posteriorly.

As the costal or lateral scutes depend, in the only important and variable part of their contour, on the form of the margins of the vertebral, it is unnecessary to describe them.

The plastron (T. XXVI) occupies about its usual relative proportion to the carapace, but it has been so much broken as to afford but little opportunity for any satisfactory or useful description. It would appear, however, from the extent of the openings for the passage of the limbs, that the animal must have possessed considerable powers of swimming, offering in this respect a very marked contrast to the testudiniform character of *E. Comptoni* and *E. testudiniformis*.

	Foot.	Inches.	Lines.
Probable total length of the carapace	1	0	0
Probable total breadth of ditto	0	10	0
Depth of the bony case	0	3	3

T. B.

Emys Delabechii. *Bell*. Tab. XXVIII.

An almost gigantic specimen of the fluviatile form of scutate Chelonia, in the collection of the Geological Survey, forms the subject of the present description. It is from the London clay of the Island of Sheppy.

This species far surpasses in size any known Emydian, whether fossil or recent; the carapace having been certainly not less than one foot nine inches in length and one foot five inches in breadth. It very clearly belongs to the form to which I have assigned it, and in some of its broader characters approximates considerably to the last species, *E. bicarinata*. The specimen is, however, unfortunately so badly injured, partly by having been originally much crushed, and partly by recent disintegration, from the decomposition of the pyrites with which it is extensively permeated, that the description must necessarily be confined to little more than general contour.

The osseous case is somewhat less deep, in proportion to its probable length and breadth, than in *E. bicarinata,* as will be seen by a comparison of their dimensions; it is consequently less sloped at the sides, which are also less curved. There is not the slightest indication of a carina, either median or lateral; but the whole vertebral region is simply flattened.

I have already had occasion to observe, that as the scutate Chelonians continue to grow, the vertebral scutes are observed to alter their form, and the relative proportion of their longitudinal and transverse diameters. This takes place particularly by the comparative abbreviation of the angular lateral projections which meet the line of junction of the margins of the corresponding costal scutes. These angles, as the animal grows, and as the scutes increase in size, become comparatively much shorter and more obtuse; and to such an extent does this take place, that in many species the sides of the vertebral scutes become very nearly parallel in old age; as may be observed in the figure of *E. bicarinata* (T. XXV), and in most recent species.

Now the specimen at present under notice, notwithstanding its great size, exhibits this indication of old age, even in a less degree than in the figured specimen of *E. bicarinata.* We could not, therefore, even if other distinctive characters were absent, for a moment confound them as one species.

In longitudinal dimensions the scutes in question ordinarily increase in proportion to the growth of the animal; and afford, in the examination of mutilated fossil Chelonian remains, approximating data for ascertaining the general size of the animal; the second and third vertebral scutes, taken together, being generally rather less than two fifths of the total length of the carapace.

The edge of the present specimen, and the injuries it has undergone, combine to render any satisfactory account of the vertebral series of osseous plates impossible; the nuchal and pygal plates being absent, and the neural wholly indistinguishable; and the plastron has been even more mutilated than the carapace.

The impressions of the vertebral scutes are tolerably perfect, as far as regards the second ($v2$), third ($v3$), and fourth ($v4$). The second and third are about as broad as they are long, irregularly hexagonal, and the lateral angles are but moderately produced; the third has the posterior margin shorter than the anterior; the fourth is rather longer than it is broad, and notably narrowed posteriorly.

The plastron exhibits at least the usual expanse of form which belongs to the typical *Emydes,* but its condition is such as to preclude any detailed description.

	Feet.	Inches.	Lines.
Probable length of the carapace	1	9	0
Probable breadth of ditto	1	5	0
Depth	0	4	8

Such are the meagre details to which we are restricted in describing by far the largest of all the fossil species of this genus. I have the gratification of offering it by name to my distinguished friend Sir Henry De la Bèche, through whose kindness I have the opportunity of including it in the present Monograph.

T. B.

FRAGMENTARY REMAINS OF EMYDIANS.

EMYS CRASSUS. Tab. XXVII.

From several such specimens kindly transmitted to me by the Marchioness of Hastings, I have selected for the subjects of Tab. XXVII two portions of a plastron; viz. the hyosternal (figs. 1, 1′) and the hyposternal (figs. 2, 2′). They are chiefly remarkable for their thickness (fig. 3), and also for their size in other dimensions.

The hyosternal shows on its outer surface (fig. 1) very strong impressions of the interspace or union between the humeral and pectoral scutes, and between the pectoral and abdominal scutes. The hyposternal shows the same kind of impression between the abdominal and femoral scutes.

These specimens were discovered in the Eocene sand at Hordwell, and are in the museum of the Marchioness of Hastings.

In Tab. XXIV, figs. 1—5, are figured some portions of the carapace of an *Emys*, from the Eocene deposits on the north shore of the Isle of Wight. These also form part of the collection of the Marchioness of Hastings.

PRINTED BY C. AND J. ADLARD, BARTHOLOMEW CLOSE.

TAB. I.

Skull of *Chelone breviceps,* nat. size.

Fig.
1. Side view.
2. Top view.
3. Base view.
4. Back view.

The following numerals indicate the same bones in each figure.

1. Basioccipital.
2. Exoccipital.
3. Supraoccipital.
4. Paroccipital.
5. Basisphenoid.
7. Parietal.
8. Mastoid.
11. Frontal.
12. Postfrontal.
14. Prefrontal (with connate Nasal and Lachrymal).
21. Maxillary.
26. Malar.
27. Squamosal.
28. Tympanic.
29. Articular.
29. Surangular.
30. Angular.
32. Dentary.

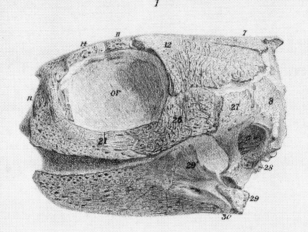

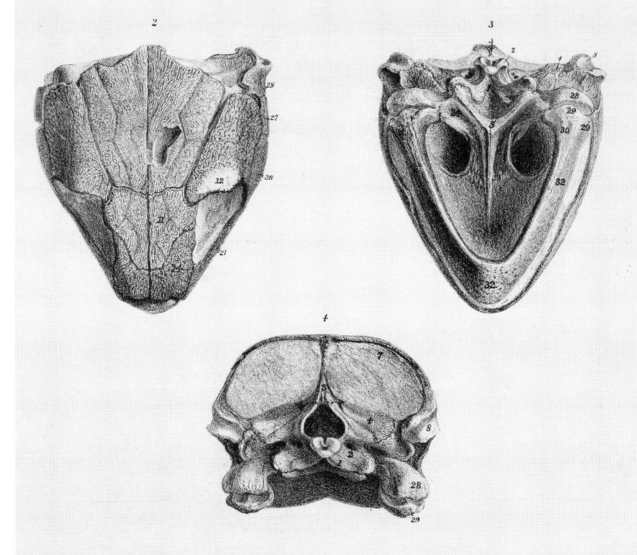

TAB. II.

Chelone breviceps, four fifths of the natural size.

Fig.

1. Upper view of the carapace.

    ch. The nuchal plate.

    $s1$—$s10$. The neural plates, most of which are connate with the neural spines.

    $pl1$—$pl8$. The costal plates, connate with the pleurapophyses or vertebral ribs.

    $v1$—$v4$. Impressions of the vertebral scutes.

    $m1$. The first marginal plate.

2. Under view of the plastron and bent-down skull.

    es. Episternal element of plastron.

    hh. Hyosternals.

    ps. Hyposternals.

    xs. Xiphisternals.

3. Side view of skull.

    7. Parietal.

    8. Mastoid.

    11. Frontal.

    12. Postfrontal.

    21. Maxillary.

    26. Malar.

    27. Squamosal.

    28. Tympanic.

4. Outline of the transverse section of the middle of the thoracic-abdominal case.

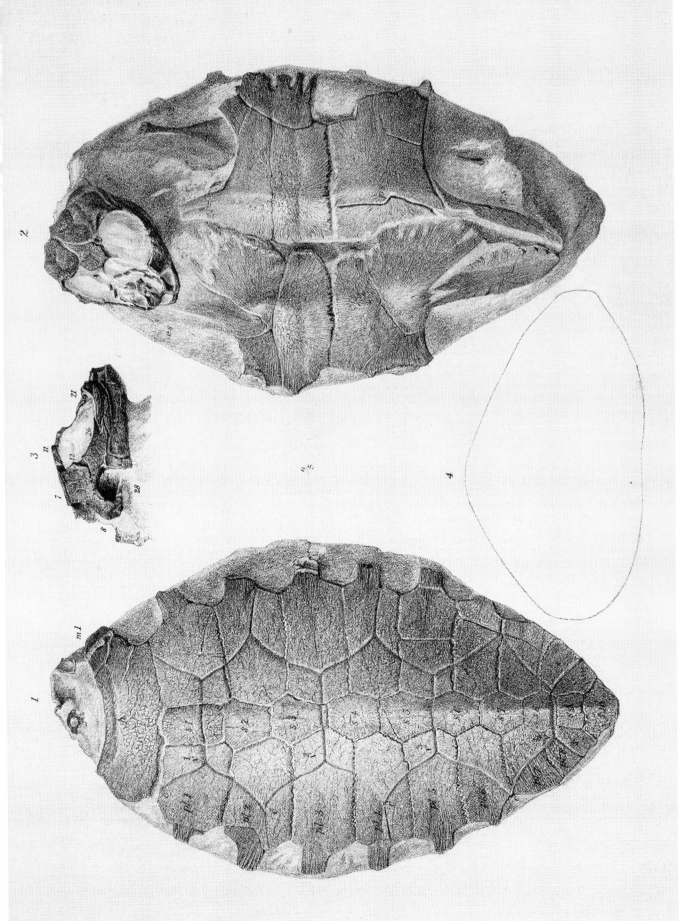

Dinkel, del et lith.

Day & Son lith. to the Queen.

TAB. III.

Skull of *Chelone longiceps*, nat. size.

Fig.
1. Side view.
2. Top view.
3. Base view.
4. Back view.

The same numerals indicate the same bones as in Tab. I.

13, 13. Vomer.
20. Palatine.
22. Premaxillary.
24. Pterygoid.

The following perishable parts are indicated by the impressions left on the bone:

fr. The frontal scute.
sy. The sincipital scute.
io. The inter-occipital scute.
oc. The occipital scute.
ol. The occipito-lateral scute.
pa. The parietal scute.
ob. The supraorbital scute.
fn. The frontonasal scute.

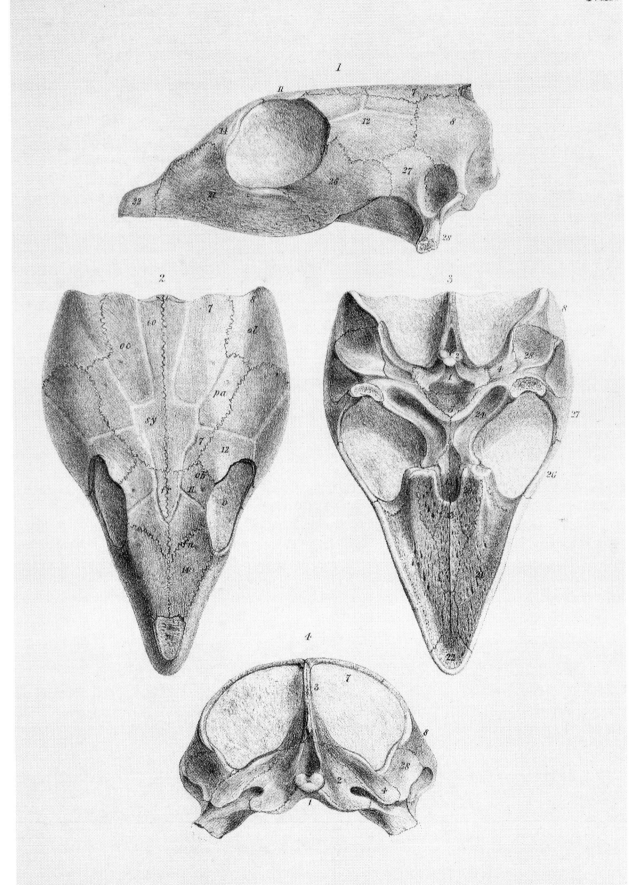

TAB. IV.

Chelone longiceps, nat. size.

Fig.
1. Skull and portion of carapace.
    *s*2. Second neural plate.
    *s*3. Third neural plate.

2. Portion of carapace.
    *s*4—*s*11. Fourth to eleventh neural plates.
    *pl*5—*pl*8. Impressions or remains of costal plates.
    *v*4—*v*5. Impressions of fourth and fifth vertebral scutes.

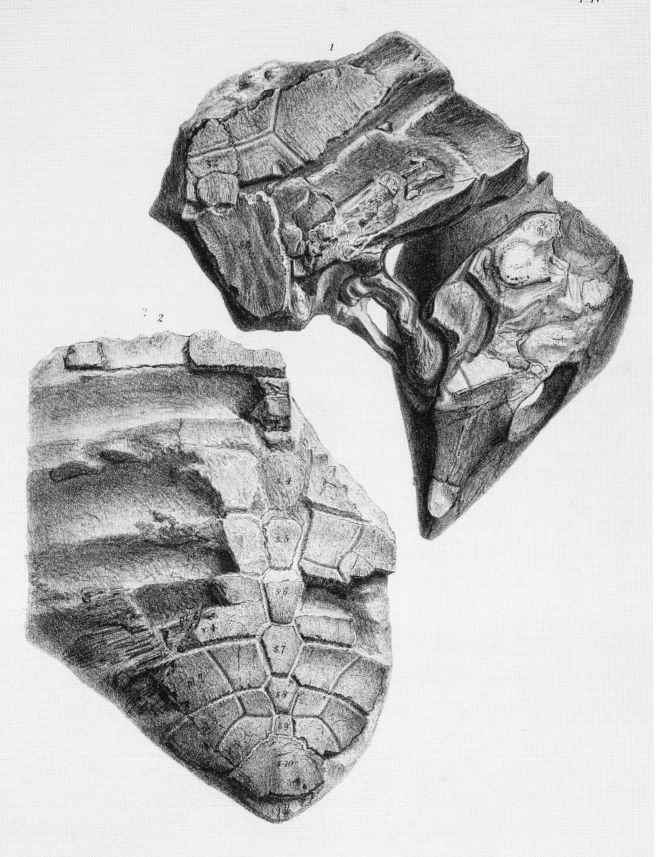

Deakel des et lith.

TAB. V

Chelone longiceps.

Fig.

1. Upper view of the carapace, two thirds of the natural size.
    ch. Nuchal plates.
    *s*1—*s*10. Neural plates.
    *pl*1—*pl*8. Costal plates.
    *v*1—*v*4. Vertebral scutes.

2. Under view of the plastron of another specimen, two thirds nat. size.
    s. Entosternal.
    es. Episternals.
    hs. Hyosternals.
    ps. Hyposternals.
    xs. Xiphisternals.
    63. The ischium.

2

1

2/3 Nat. size.

TAB. VI.

Chelone latiscutata, nat. size.

Fig.

1. Upper surface of the carapace.

    $s1$. First neural plate. $pl1$. First costal plate.

    $s2$. Second do. $pl2$. Second do.

    $s3$. Third do. $pl3$. Third do.

    $s4$. Fourth do. $pl4$. Fourth do.

    $s5$. Fifth do. $pl5$. Fifth do.

    $s6$. Sixth do. $pl6$. Sixth do.

    $v2$. The impression of the second vertebral scute.

    $v3$. The impression of the third vertebral scute.

2. Lower surface of the plastron.

    s. Extremity of the entosternal.

    hs. The hyosternal.

    ps. The hyposternal.

    xs. Impressions of the xiphisternals.

3. Anterior surface of the carapace and plastron, showing the degree of convexity of those parts, and the depth of the osseous case.

    51 is the scapular arch.

TAB. VII.

Chelone convexa.

Fig.

1. Upper view of the carapace, nat. size.

2. Under view of the plastron, nat. size.
   The letters and figures indicate the same parts as in the previous subjects.
   53. The humerus.   64. The pubis.   65. The femur.

3. Under view of the symphysis and part of the rami of the lower jaw.

4. Outline of the curve of the carapace.

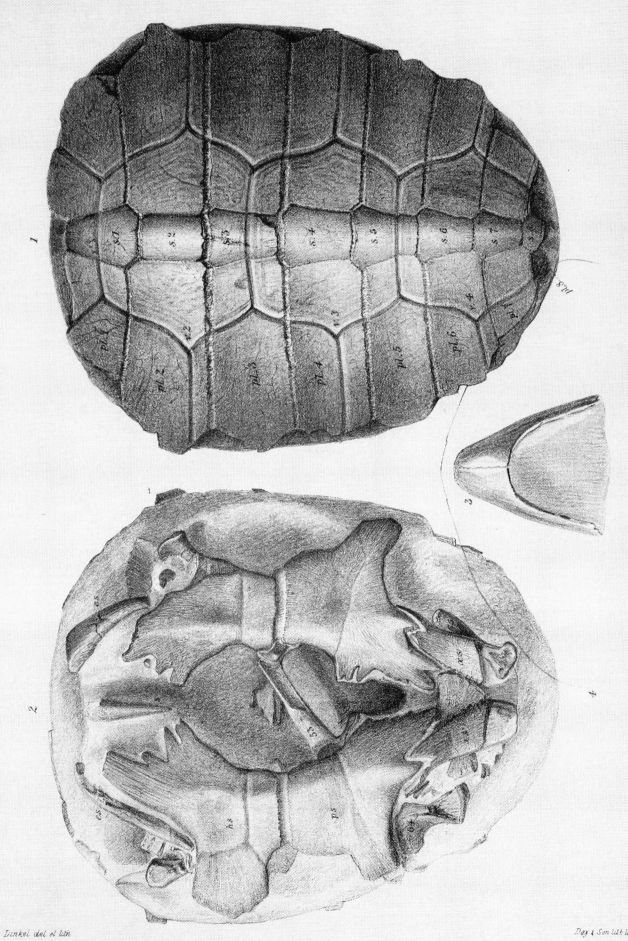

TAB. VIII.

Chelone subcristata.

Fig.
1. Upper view of the carapace, three fifths of the natural size.

2. Under view of the plastron of the same specimen.

The letters and figures indicate the same parts as in the preceding subjects.

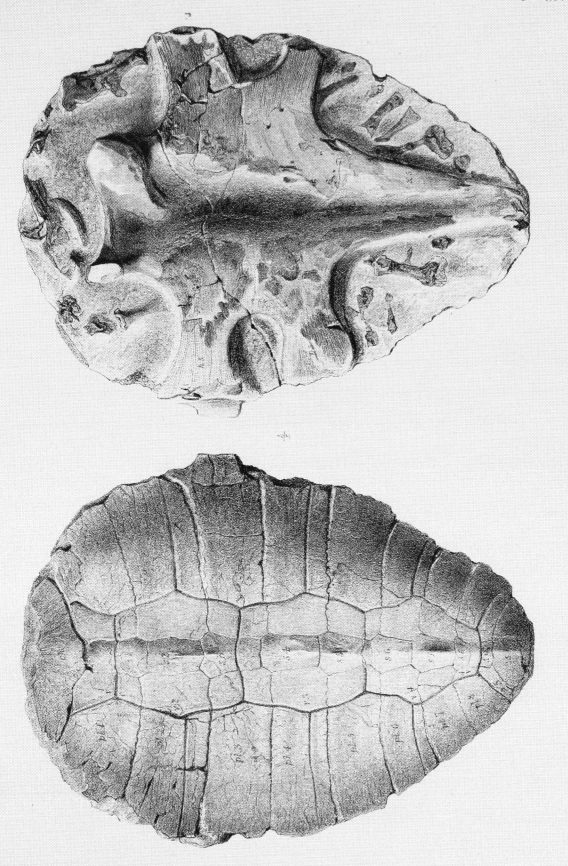

TAB. VIII*A*.

Chelone subcarinata.

Fig.
1. Upper view of the carapace, seven twelfths of the natural size.

2. Under view of the plastron of the same specimen.

3. Outline of the transverse contour of the carapace.

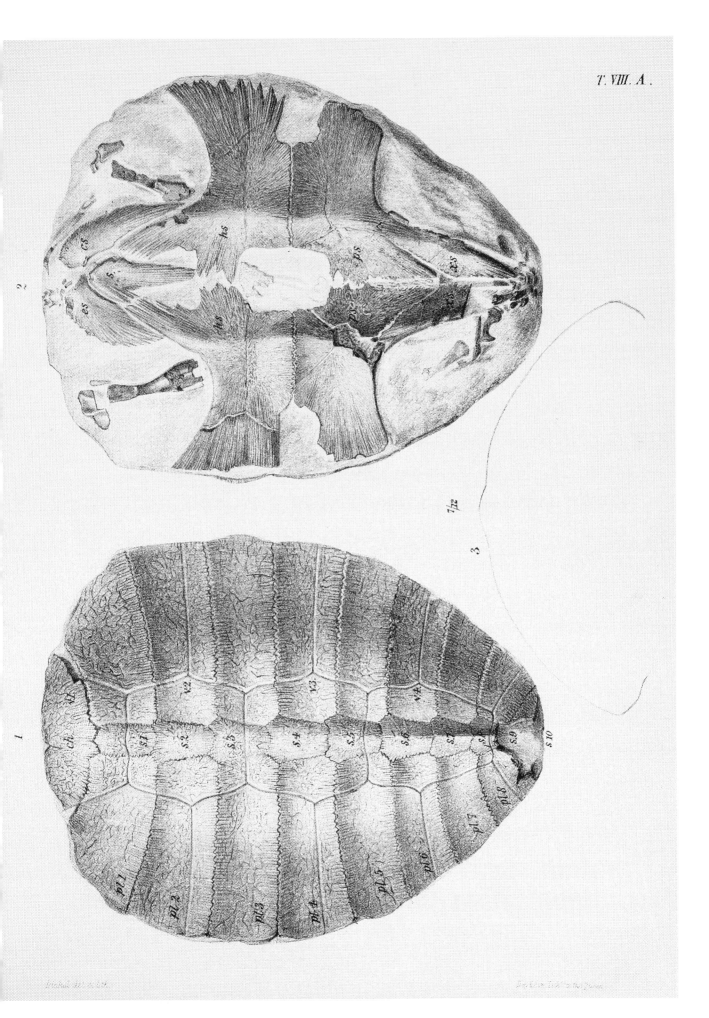

TAB. IX.

Skull of *Chelone planimentum,* nat. size.

Fig.

1. Under view of the lower jaw, showing the characteristic extent and flatness of the symphysis.

2. Upper view of the skull.

3. Side view of the skull.

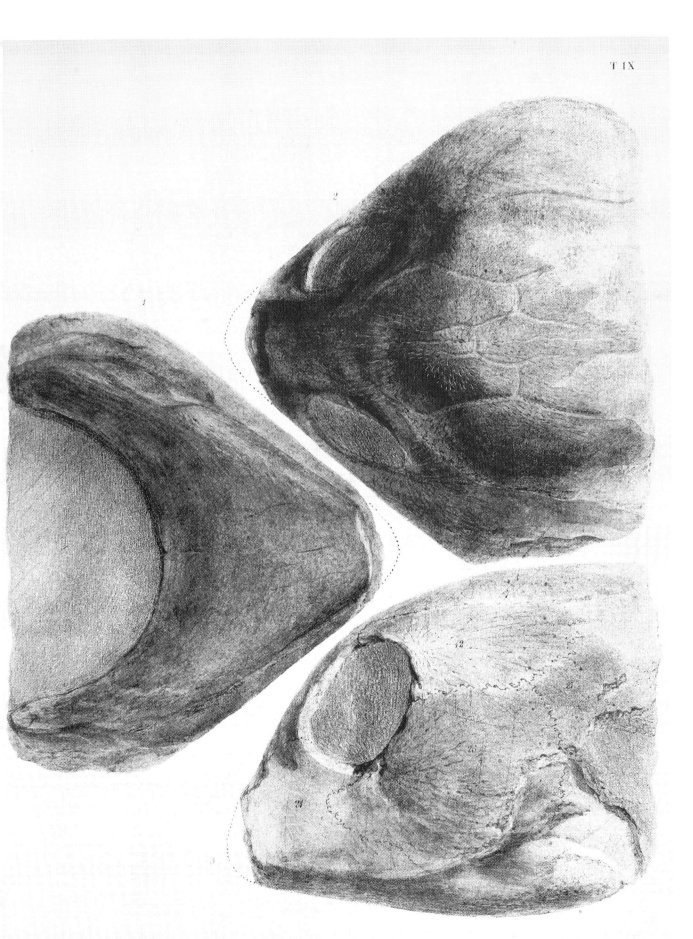

TAB. X.

Chelone planimentum.

Under or inside view of the carapace, half the natural size.

$s5$—$s8$. The fifth to the eighth neural plates which are anchylosed to the neural spines.

$s9$—$s11$. Three succeeding neural plates which do not become so anchylosed.

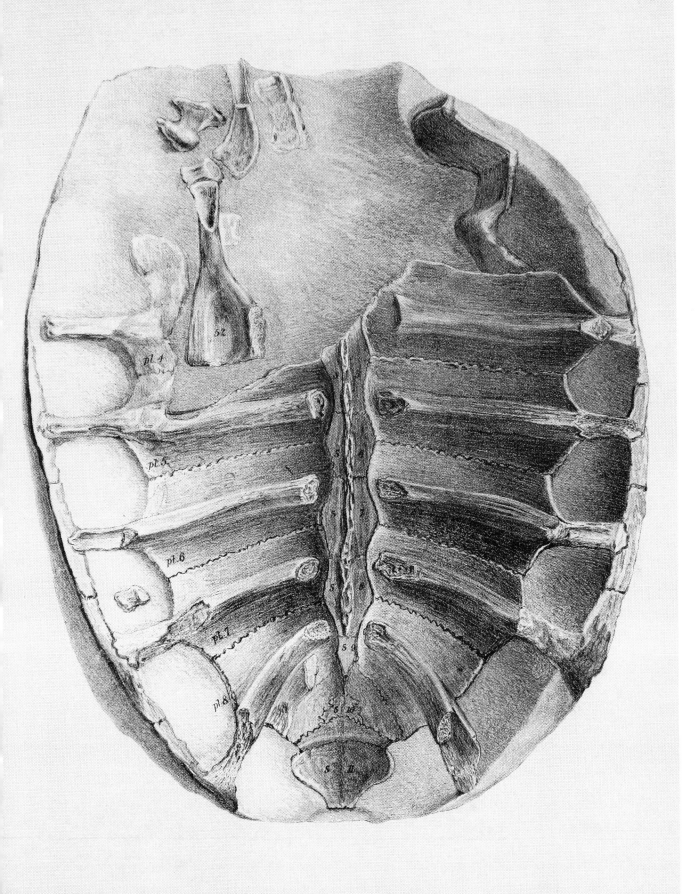

TAB. X*A*.

Chelone planimentum.

View of a cast of the under or inner surface of the carapace; with portions of the ribs and marginal plates: one third the natural size.

*pl*1—*pl*8. Impressions of the costal plates and ribs.

*m*11, *m*12. The eleventh and twelfth marginal plates.

½ Nat. Size.

TAB. XI.

Skull of the *Chelone crassicostata*, two thirds of the nat. size.

Fig.
1. Upper view, showing chiefly a cast of the under surface of the bones.
   12. The situation of the postfrontals.   14. The prefrontals.   15. The nasals.

2. Side view.

3. Under view of the symphysis menti.

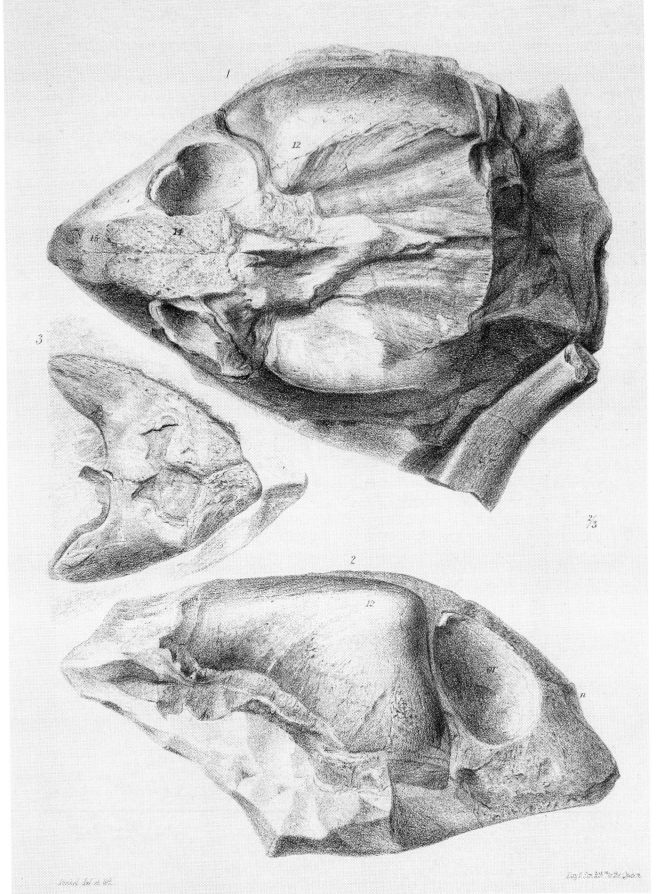

TAB. XII.

Chelone crassicostata.

View of the inner surface of the carapace, half the natural size.

ch. The nuchal plate.

*pl*1. The first costal plate and adherent rib.

*pl*2. The second ditto.

*pl*3. The third ditto.

*pl*4. The fourth ditto.

*pl*5. The fifth ditto.

*pl*6. The sixth ditto.

*pl*7. The seventh ditto.

*pl*8. The eighth ditto. The figures are placed above the free projecting ends of the ribs; they belong to the second and the ninth dorsal vertebræ inclusive.

*×*9. The ninth neural plate.

*×*10. The tenth neural plate.

py. The last neural or the " pygal" plate.

These are developed, like the nuchal plate, in the substance of the derm, and do not become confluent with any parts of the endoskeleton.

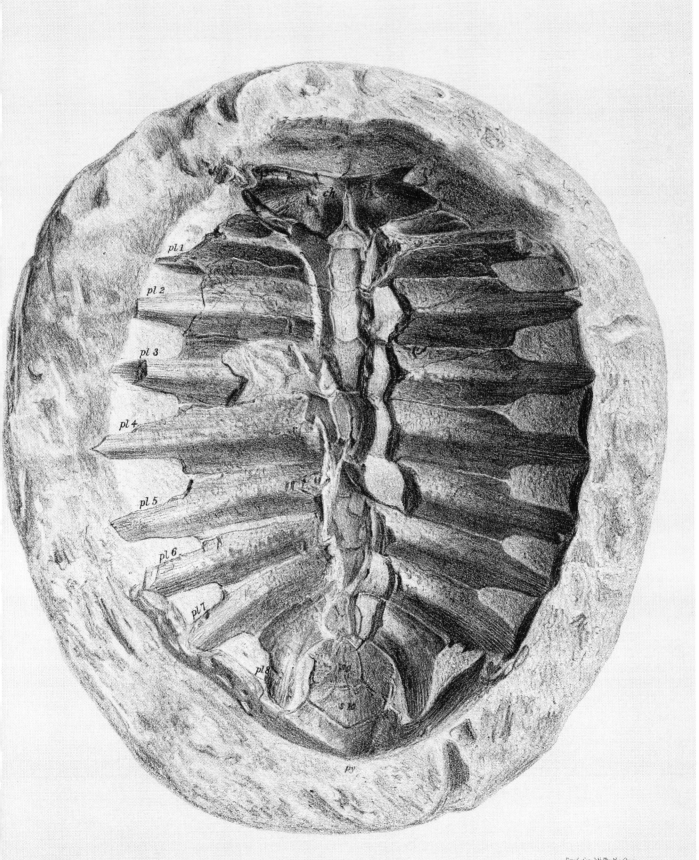

pl 1

pl 2

pl 3

pl 4

pl 5

pl 6

pl 7

pl 8

TAB. XIII.

Chelone crassicostata.

Fig.
1. Portion of the plastron seen from the under side.

  hs. The hyosternal; *ps*, the hyposternal; *xs*, the xiphisternal.   52 is the coracoid, exposed by the removal of the right hyosternal.

2. Cast in the matrix of a portion of the carapace of the same individual, showing the characteristic thickness of the ribs, *pl, pl.*

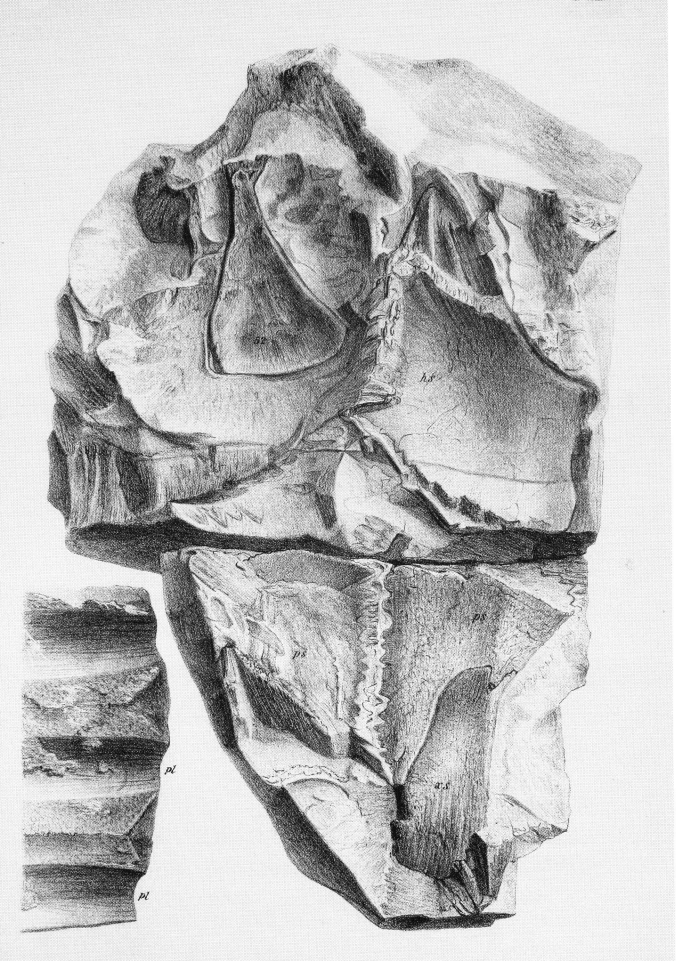

TAB. XIII*A*.

Chelone crassicostata.

Portion of the carapace, half the nat. size, with the corresponding series of marginal plates. $m6$—$m12$, and py.

$pl3$ and $pl4$, the third and fourth costal plates of the left side; $pl5$—$pl8$, impressions in the matrix of the succeeding costal plates.

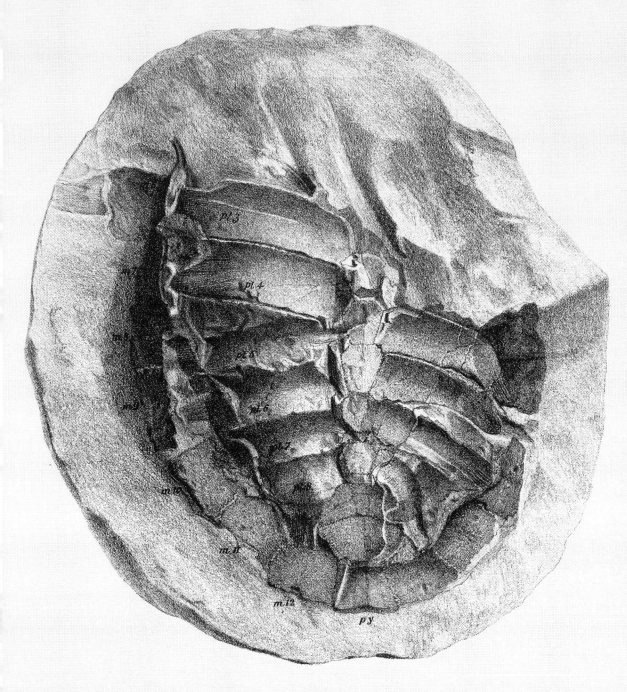

½ Nat. Size

TAB. XIII*B*.

Chelone crassicostata, half the nat. size.

Fig.

1. Outside view of the carapace. $s7$—$s10$, the seventh to the tenth neural plates inclusive: $pl2$—$pl8$, the second to the eighth costal plates of the left side ; $pl3$—$pl8$, the third to the eighth, inclusive of the right side.

2. The plastron dislocated and crushed in, with part of the skull showing the orbit o ; hs, the hyosternal ; ps, the hyposternal ; xs, the xiphisternal ; 51, the scapula : 52, the coracoid.

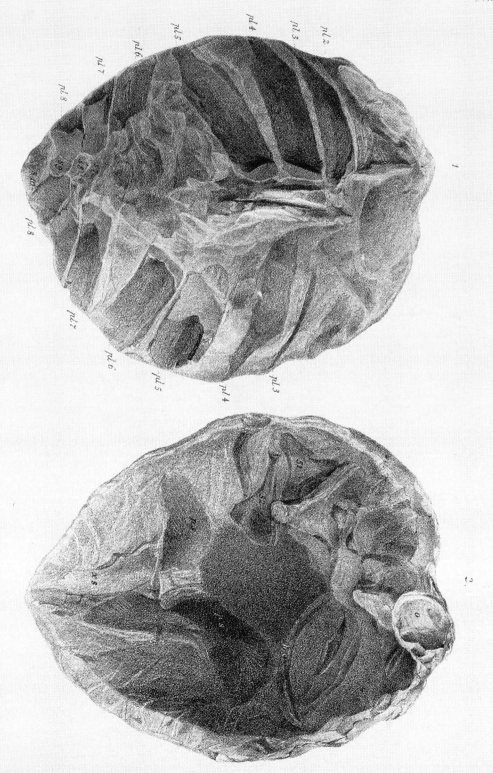

T. XII. B.

From Nature & on Stone by J. Erxleben.

Roy & Son Lith.

TAB. XIV.

Chelone declivis.

Fig.
1. Upper view of the left moiety of the carapace, nat. size.
   - *ch.* Impression of the nuchal plate.

$s1$. First neural plate.	$pl1$. First costal plate.
$s2$. Second ditto.	$pl2$. Second ditto.
$s3$. Third ditto.	$pl3$. Third ditto.
$s4$. Fourth ditto.	$pl4$. Fourth ditto.
$s5$. Fifth ditto.	$pl5$. Fifth ditto.
$s6$. Sixth ditto.	$pl6$. Sixth ditto.
$s7$. Seventh ditto.	$pl7$. Seventh ditto.

   - $v1$. Impression of the first vertebral scute.
   - $v2$. Impression of the second vertebral scute.
   - $v3$. Impression of the third vertebral scute.
   - $v4$. Impression of the fourth vertebral scute.

2. View of part of the dislocated right half of the carapace. The same letters and numerals signify the same parts.

5. Outline of the natural transverse curve of the carapace.

TAB. XV.

Skull of *Chelone cuneiceps*, nat. size.

Fig.
1. Side view.

2. Top view.

3. Base view.

4. Back view.

The letters and numerals indicate the same parts as in Tabs. I and III.

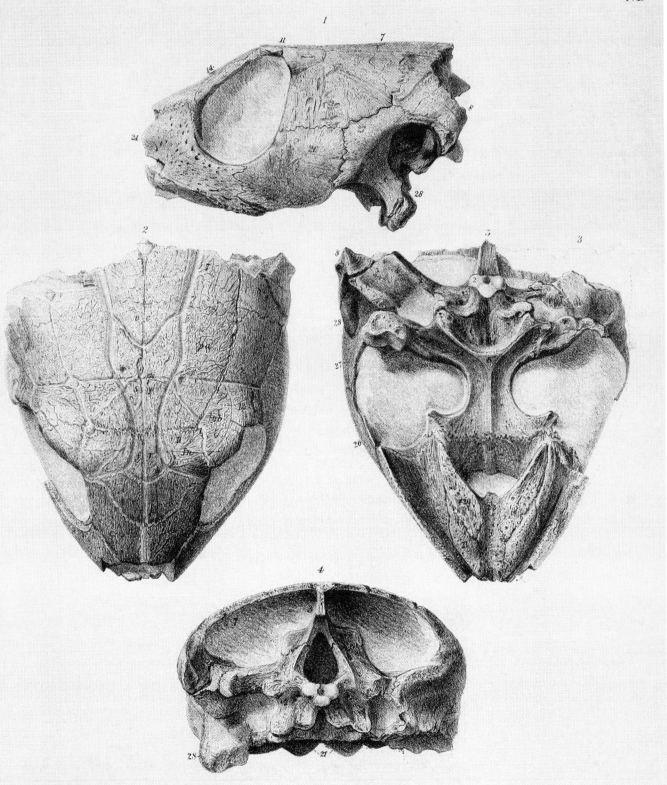

TAB. XVI.

Trionyx Henrici.

Fig.

1. Upper view of the carapace, wanting the nuchal plate, half nat. size.
   s_1—s_7. The first to the seventh neural plates inclusive.
   pl_1—pl_8. The eight costal plates of the left side.

2. Outline of the transverse curvature of the upper surface of the carapace.

3. The nuchal plate of probably the same species of *Trionyx ;* but of another individual : half the natural size.

In Zinc by J. Erxleben.

TAB. XVI*A*.

The carapace of the *Trionyx Barbaræ*, half nat. size.

a, Shows the longitudinal contour of the middle of the upper surface, and fig. 2, the transverse curvature of the carapace.

TAB. XVII.

Trionyx incrassatus.

Fig.

1. Inside view of the carapace, wanting the nuchal plate, half nat. size.

   s_1—s_7. The first to the seventh neural plates, with the connate neural spines and arches.

   pl_1—pl_8. The eight costal plates of the right side.

   c_1. The place of attachment of the outer end of the first dorsal rib.

   c_2—$_9$. The portions of the consecutive dorsal ribs that become connate with the costal plates.

2. Transverse contour of the upper surface of the carapace.

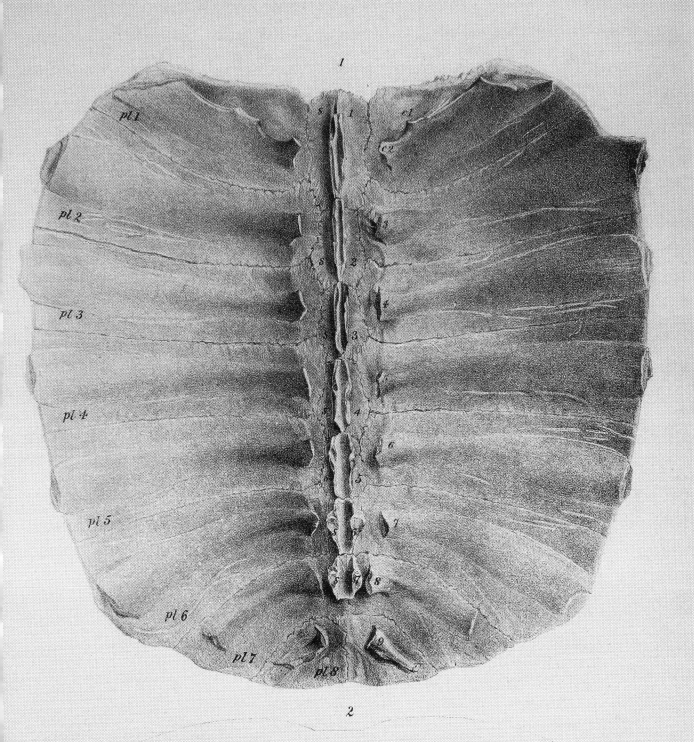

TAB. XVIII.

Trionyx incrassatus.

Fig.

1. Upper or outside view of the fore part of the carapace with the nuchal plate, half nat. size.

   ch. The nuchal plate.

   s_1. The first neural plate.

   s_2. The second neural plate.

   $pl_1—pl_3$. The three anterior costal plates of the left side.

2. Under or inside view of the same specimen.

   ch. The nuchal plate.

   s_1. The first neural plate connate with the neural-spine of the second dorsal vertebra, of which part of the under surface of the centrum is here preserved and shown.

   s_2. The second neural plate, with the connate neural arch of the third dorsal vertebra.

   $pl_1—pl_3$. The three anterior costal plates of the right side.

   c_1. The place of attachment of the outer end of the first dorsal rib.

   c_2. The portion of the second dorsal rib that becomes connate with the first costal plate. c_2'. The extremity of the rib that again becomes free.

   c_3. The third rib attached to the second costal plate.

   c_4. The fourth rib attached to the third costal plate.

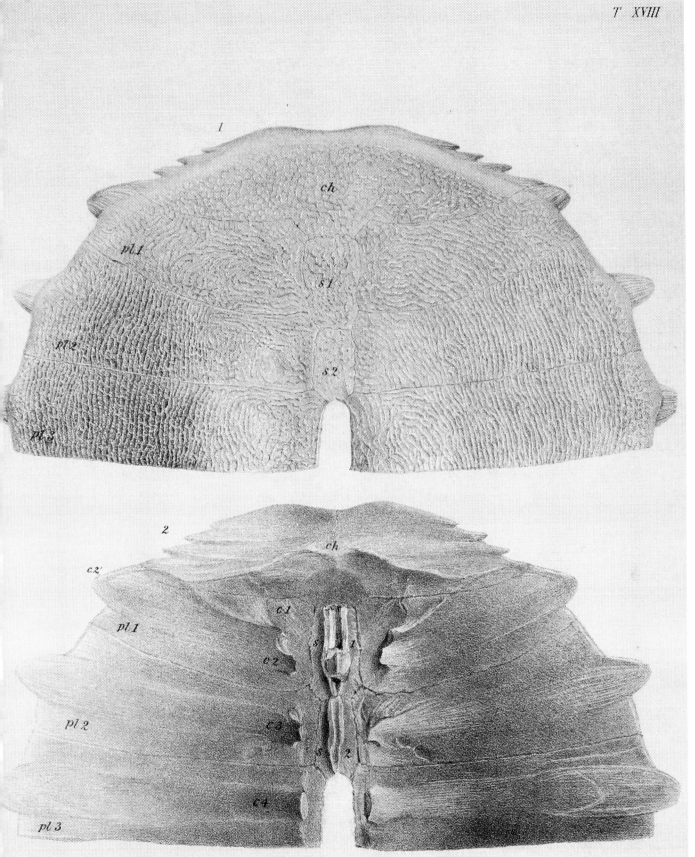

TAB. XVIII*A*.

Trionyx rivosus.

Fig.
1. Outside view of hinder half of the carapace, nat. size.
   pl_4—pl_8. The fourth to the eighth costal plates inclusive.

2. Inside view of the same specimen.
   pl_4—pl_8. The fourth to the eighth costal plates inclusive.
   5—9. The ribs (fifth to the ninth dorsal inclusive) connate with the above costal plates.
   s_6. The neural arch confluent with the sixth neural plate.

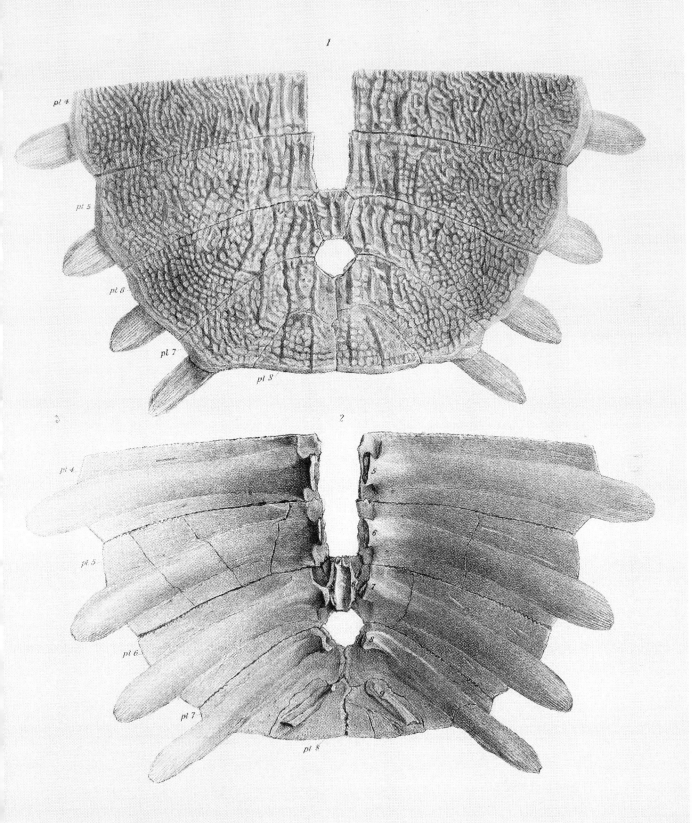

TAB. XIX.

Trionyx incrassatus.

The parts are figured half the nat. size.

Fig.

1. Entosternal.
2. Episternal, under side.
2'.   Ditto.   upper side.
3. Part of right hyosternal, and the hyposternal, under or outer side.
3'. *hs,*   Ditto,   and *ps,*   ditto,   upper side.
4. Two views of the scapula (51), and connate acromial clavicle (58) ; 4″ shows the osseous structure of the end of the scapula, nat. size.
5. The coracoid.
6, 6'. Two views of the ilium.
7, 7'. Two views of the femur.
8, 8. Two views of an ungual phalanx, nat. size.
9. Under view of the sixth cervical vertebra ; *y, y,* the hypapophyses.
9'. Upper view of the same vertebra ; *c,* the anterior convex articular surface of the centrum ; *n,* the neural arch ; *z, z,* the zygapophyses (oblique or articular processes).
9″. Back view of the same vertebra, showing the double concavity of the articular surface.

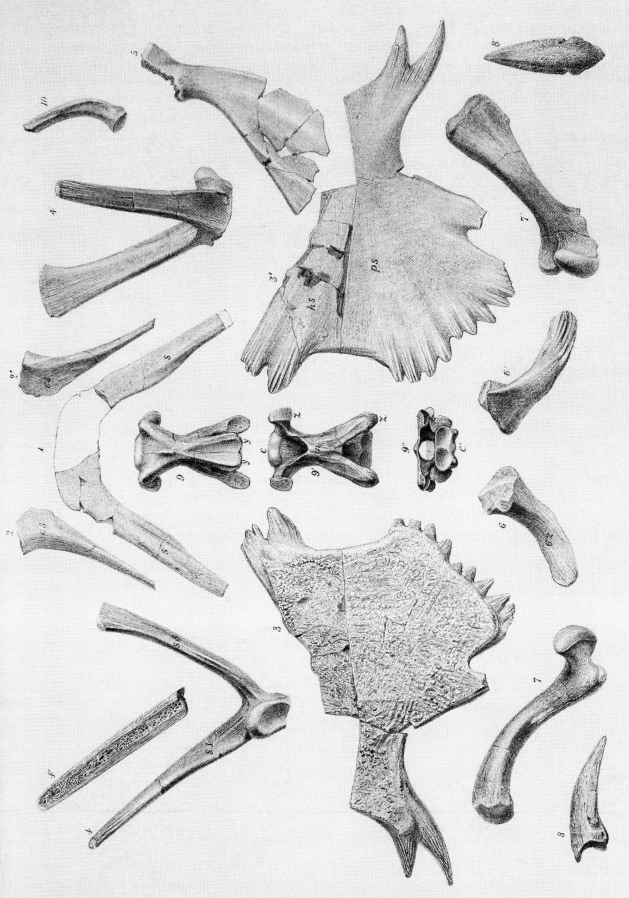

½ Nat. size.

Jac Dinkel del et lith.

TAB. XIX+.

Trionyx marginatus.

Fig.

1. Upper view of the carapace, wanting the nuchal plate, half the nat. size. The letters and figures indicate the same parts as in Tab. XVI.

2. Outline of the transverse curvatures of the upper surface of the carapace.

TAB. XIX*B*.

Fig.

1. Upper surface of the third costal plate, right side, of the *Trionyx circumsulcatus,* nat. size.
2. Articular margin of ditto ; showing the natural curve and thickness of the plate.
3. Peripheral margin of ditto ; showing the groove.
4. Upper surface of the third costal plate, right side, of the *Trionyx marginatus.*
5. Sutural margin of ditto.
6. Peripheral margin of ditto.
7. Upper surface of a fragment of a costal plate of the *Trionyx pustulatus,* nat. size.
8. Under surface of ditto, showing the large and prominent adherent rib.
9. Peripheral margin of ditto.

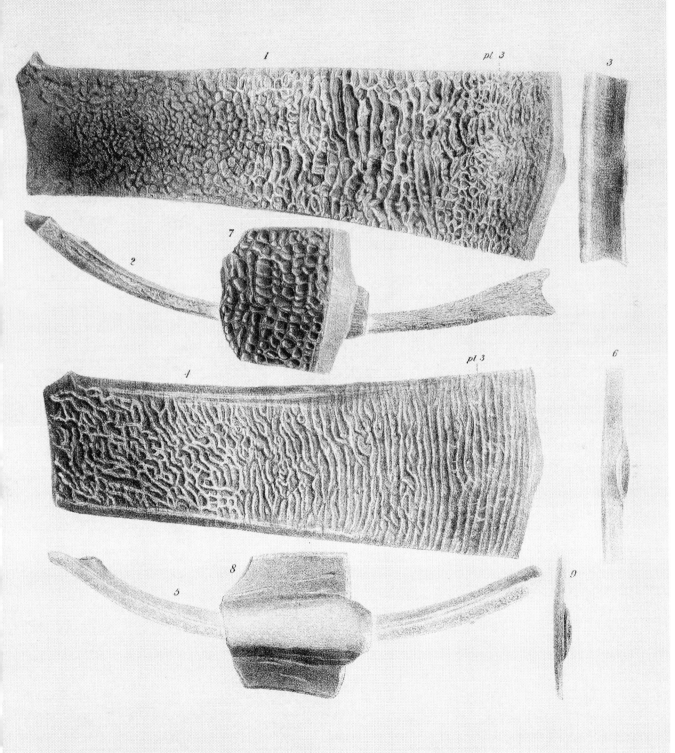

TAB. XIX*C.*

Trionyx planus.

Fig.
1. Upper view of the hind part of the carapace, half the nat. size.

2. Inside view of the same specimen; *pl*5—*pl*8, the fifth to the eighth costal plates inclusive.

3. Peripheral border of the fifth costal plate, nat. size.

4. Peripheral border of the fifth costal plate of the *Trionyx Barbaræ,* showing the difference in their relative thickness.

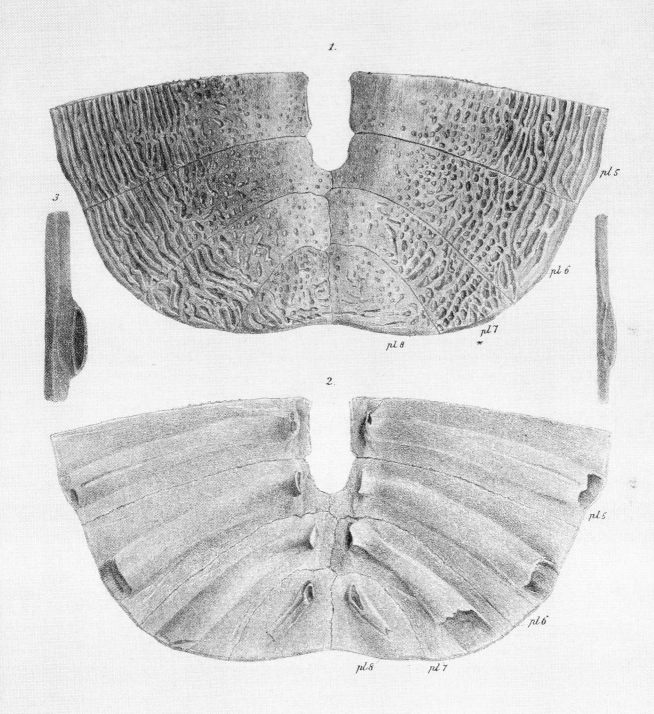

From Nature & on Stone by J. Erxleben

Day & Son, Lith Printers, London

TAB. XIX*D*.

Fig.

1. Under view of the lower jaw of a turtle (*Chelone*), nat. size. (Hordwell Cliff).

2. Upper view of the same specimen.

3. Upper view of the fourth cervical vertebra of a *Trionyx*; *d*, diapophysis. Nat. size.

4. Side view of the same specimen (Hordwell Cliff).

5. The left os pubis of a *Trionyx*, nat. size (Hordwell Cliff).

6. Part of the plastron of *Trionyx planus*, half the nat. size; *hs*, hyosternal; *ps*, hyposternal.

7. Right hyposternal of a *Trionyx*, from Bracklesham, half the nat. size. (In the Collection of Frederick Dixon, Esq., F.G.S.)

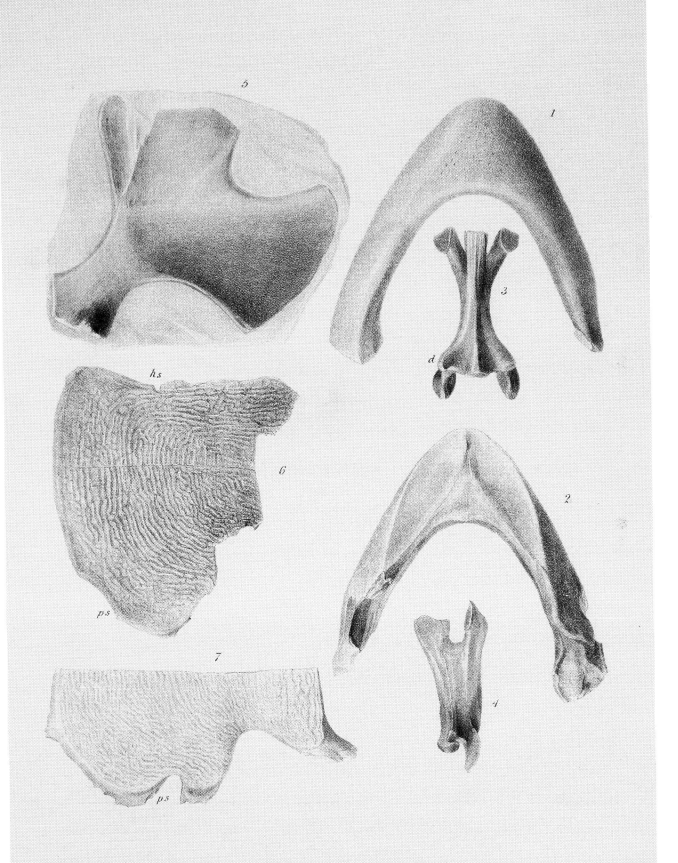

On Stone by J. Erxleben.

Day & Son lith. to the Queen.

TAB. XX.

Emys Comptoni, nat. size.

Fig.
1. Upper view of the carapace.

2. Under view of the plastron.

3. Side view, showing the marginal plates, m_4—m_7.

4. Front view.

5. Back view.

TAB. XXI.

Platemys Bullockii, half the nat. size.

The plastron and some of the marginal plates, *a a.*

    s. Entosternal.

    es. Episternals.

    hs. Hyosternals.

    ps. Hyposternals.

    xs. Xiphisternals.

    ig. Intergular scute.

    gu. Gular scutes.

    hu. Humeral scutes.

    pe. Pectoral scutes.

    ab. Abdominal scutes. The impression between these and the pectoral scutes crosses the intercalated supernumerary bones between the hyosternals and hyposternals.

    fe. Femoral scutes.

    an. Anal scute.

Where the bones or scutes are in pairs, the figures or letters are placed on one of each pair.

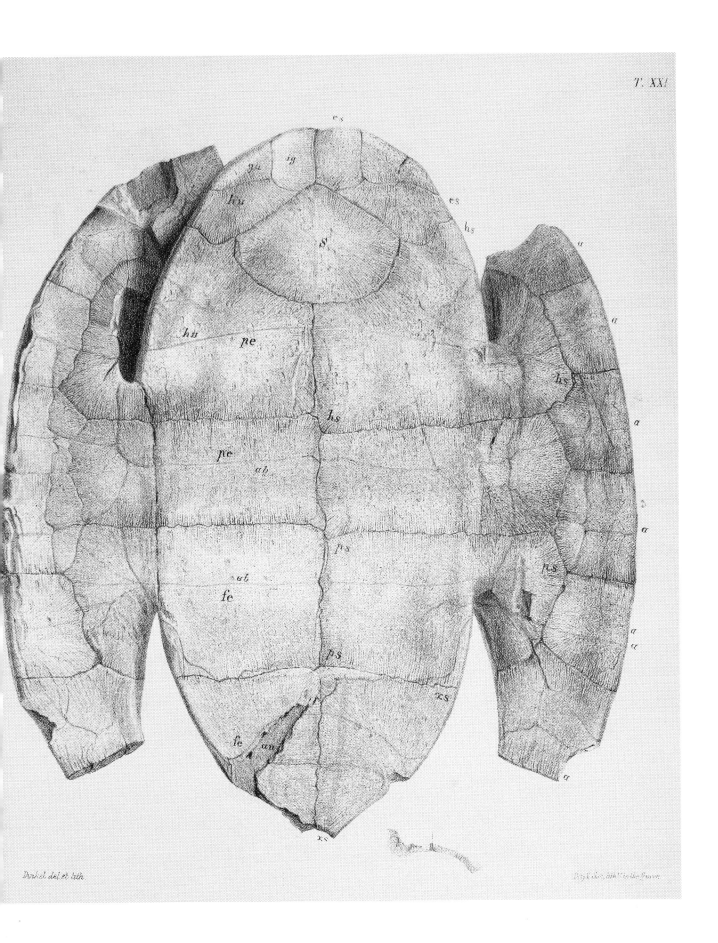

TAB. XXII.

Emys lævis, two thirds the nat. size.

Fig.

1. Upper view of carapace.

2. Under view of plastron.

3. Side view

4. Front view, showing the curvature and depth of the specimen.

The letters and figures signify the same parts as in the preceding figures; *hp,* the accessory pieces of the plastron.

TAB. XXIII.

Platemys Bowerbankii, half the nat. size.

Fig.
1. The mutilated carapace.

2. The plastron.

The letters and figures indicate the same parts as in the preceding Table.

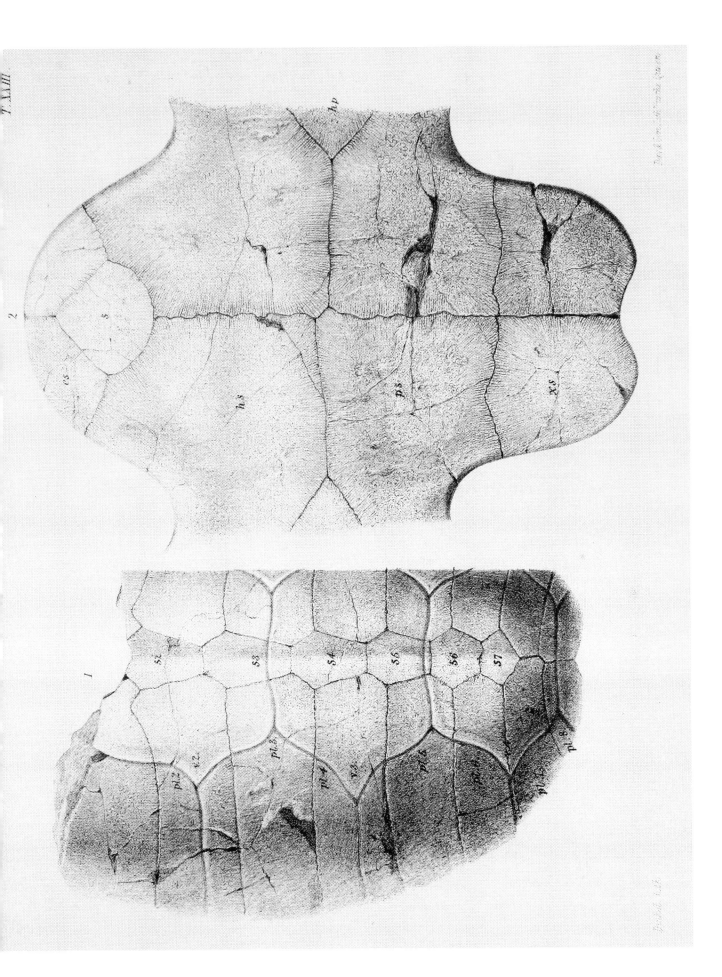

TAB. XXIV.

Emys testudiniformis, nat. size.

Fig.

1. Nuchal plate.

2. Upper surface of the third neural plate.

3. Under surface of ditto.

4. Upper surface of the fifth neural plate.

5. A posterior marginal plate.

6. Front view of a mutilated cuirass in which the carapace has been slightly depressed; the natural curve, across the middle, is indicated by the outline.

T.XXIV

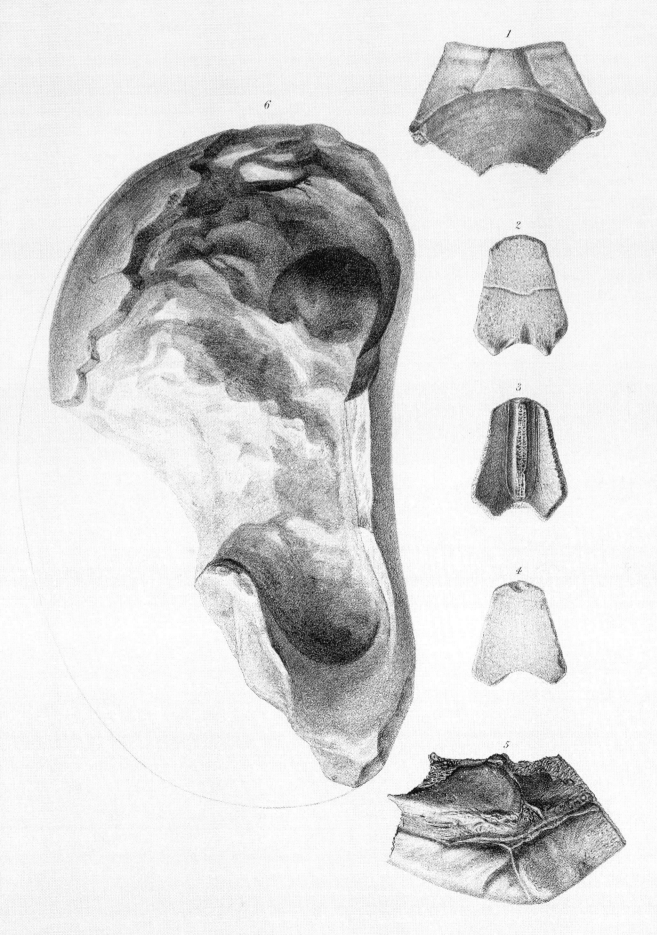

Pinkel. del. et lith.

Day & Son, lith.rs to the Queen.

TAB. XXV.

Emys bicarinata, two fifths the nat. size.

The letters and numbers on the carapace, the upper surface of which is figured, indicate the same parts as in the previous figures.

The transverse contour of the carapace is given in outline above.

⅔ Nat. Size

TAB. XXVI.

Emys bicarinata, two fifths the nat. size.

The plastron; *s,* the entosternal piece.
xs, the xiphisternals.

⅔ *Nat.Size*

TAB. XXVII.

Emys crassus, two thirds the nat. size.

Fig.
1. Outside view of the left hyposternal.

1′. Inside view of ditto.

2. Outside view of right hyosternal.

2′. Inside view of ditto.

3. Sutural border of hyosternal, nat. size.

Dinkel, del. et lith.

Day & Son, lith.rs to the Queen.

TAB. XXVIII.

Emys Delabechii, two fifths the nat. size.

Upper or outside view of the carapace; the transverse curve is given in the outline above.

pl_1—pl_8. The eight costal plates of the left side.

v_2—v_4. The second to the fourth vertebral scutes inclusive.

2/5 Nat. Size

MONOGRAPH

ON

THE FOSSIL REPTILIA

OF THE

LONDON CLAY, AND OF THE BRACKLESHAM AND OTHER TERTIARY BEDS.

PART I. SUPPLEMENT No. 1.
PAGES 77—79; PLATES XXVIIIA—XXVIIIB.

CHELONIA (EMYS).

BY

PROFESSOR OWEN, D.C.L., F.R.S., F.L.S., F.G.S., &c.

Issued in the Volume for the Year 1856.

LONDON:
PRINTED FOR THE PALÆONTOGRAPHICAL SOCIETY.
1858.

SUPPLEMENT

TO THE

EOCENE CHELONIA.

EMYS CONYBEARII, *Owen.*   T. XIII and XIV.

The subject of the present description is the most complete specimen of fossil fresh-water Tortoise (*Emys*) which has hitherto come under my observation.   It was obtained, like the *Emys Delabechii*, Bell, which it most resembles, from the Eocene clay of Sheppey Island, and forms part of the large and instructive collection of Sheppey fossils belonging to J. S. Bowerbank, Esq., F.R.S.   It consists of both carapace (T. XXVIII *A*), and plastron (T. XXVIII *B*), giving the natural curves, depth, and periphery of the portable abode of the animal.   Every constituent bone of the complex roof of this abode is preserved, and the impressions of every horny scute can be traced, uninterruptedly, upon its uninjured surface.   The floor or "plastron" is, unfortunately, not so entire, but its margins are unbroken, exhibiting the characteristic contour.

The dimensions of this noble specimen of Eocene *Emys* fall little short of those of the *Emys Delabechii*, the length of the carapace being 1 foot $6\frac{1}{2}$ inches, and its breadth 1 foot 3 inches.

The forms and proportions of the nuchal (*ch*), pygal (*py*), and their connecting series of median neural plates (*s* 1 to *s* 9), are accurately shown in T. XXVIII *A*: the eighth neural plate is obliterated, or has its place taken by the extension of ossification of the seventh and eighth costal plates towards the median line.   The abraded and fractured state of the carapace of *Emys Delabechii** does not permit of a comparison of this particular.

There is no trace of a median elevation or keel in *Emys Conybearii*: the neural plates in *Emys bicarinata*† are broader in proportion to their length than in the present species, and the series is not interrupted.

---

* 'Monograph on Fossil Reptilia of the London Clay,' T. xxviii, Palæontographical Memoirs for 1848.

† Ib., T. xxv.

In the costal plates (T. XXVIII *A, pl* 1—*pl* 8) the general form and proportions are sufficiently clearly represented; but it is worthy of special remark that the first (*pl* 1) has a small quadrate portion marked out on its inner or median border, by the impression of a supplementary median scute, which I have not observed in other Emydians, recent or fossil; and should that supplementary scute be constant in the present species, such species might be indicated by a detached fossil first costal plate (*pl* 1).

In Emydians generally the vertebral scutes are five in number; the costal scutes are four pairs: collectively, when they have been called "discal" scutes or plates, thirteen is the number. In the present extinct species they were fourteen in number, owing to the division of the first vertebral scute (*v* 1), or to the interposition of a small transversely extended quadrate scute between it and the second vertebral scute (*v* 2). It will be fortunate if other specimens of the *Emys Conybearii* should show this character to be a constant one.

In *Emys Delabechii* the carapace is unluckily mutilated at the part requisite for determining whether the supplementary median scute existed. The second vertebral scute (*v* 2) in *Emys Delabechii* is as broad as it is long: in *Emys Conybearii* the breadth of the same scute is one third greater than its length. A similar difference of proportions, with more produced and acute lateral angles, characterises the third vertebral scute of *Emys Conybearii* as compared with *Emys Delabechii*.

The fourth vertebral scute in *Emys Conybearii* has the front half of its lateral margin wavy or crenate, not a simple sigmoid line, as in *Emys Delabechii*. The breadth and depth of the scutal impressions are alike in both species.

In *Hydromedusa* the number of discal scutes, or those inclosed by the marginal scutes, is fourteen, as in the present fossil; and, as it appears to me, by a similar transverse division of the first vertebral scute; only the dividing line crosses the scute more anteriorly, so that the proportions of the front and hind divisions are reversed. Dr. Gray regards the front division as the homologue of the front marginal scutel called "nuchal" in other *Chelydidæ* and in *Emydidæ*, and characterises *Hydromedusa* as having "the nuchal plate large, placed behind the front marginal plate, like a sixth vertebral;" but if the carapace of *Hydromedusa depressa*** be compared with that of *Chelodina oblonga*,† and *Sternothærus Derbianus*,‡ the marginal series of scutes will be seen to be similar in number in the two species, and their condition to be essentially that which is defined in the characters of *Sternothærus*, as "Nuchal plate none."§ The part called "nuchal plate" in *Hydromedusa depressa* answers to the front part of the first vertebral plate in *Sternothærus* and *Chelodina*: the different position of the dividing line in the *Emys Conybearii* more strongly marks the true homological character of the first of the median series of discal scutes. I conclude, therefore, that, as in

---

* 'Catalogue of Shield Reptiles,' Brit. Mus., 4to, p. 59.

† Ib., tab. xxvi. ‡ Ib., tab. xxii. § Ib., p. 51.

Sternothærus and *Hydromedusa*, the nuchal scute is absent in *Emys Conybearii*, the number of marginal scutes being 12—12.

In *Emys nigricans* the nuchal scute is absent, and the number of marginal scutes is as in *Emys Conybearii*.

The carapace of *Emys Conybearii* is moderately convex at the middle, and very slightly concave towards the margin, which is gently raised before and behind.

The plastron, as in the section of *Emydidæ*, including *Emys* proper,[*] is solid, truncate before and notched behind (T. XXVIII *B*), attached to the carapace by sutures of bone. The number and character of the sternal plates I have been unable to determine : there is no transverse joint as in the Box-tortoises (*Cistudo, Lutremys*).

The chief peculiarity of the plastron of the *Emys Conybearii* is the concavity of its middle three fifths (T. XXVIII *B*, fig. 2)—a modification which is rarely seen save in some true Tortoises. This character, coupled with the divided first vertebral scute, would probably be deemed worthy of supporting a sub-generic distinction by some erpetologists.

---

[*] Gray, tom. cit., *A, a,* p. 15. It is called "sternum" in this work; but the "plastron" includes many other elements of the skeleton besides the sternum; to say nothing of the bony scutes, superadded to the sternum and abdominal ribs.

py

1/2

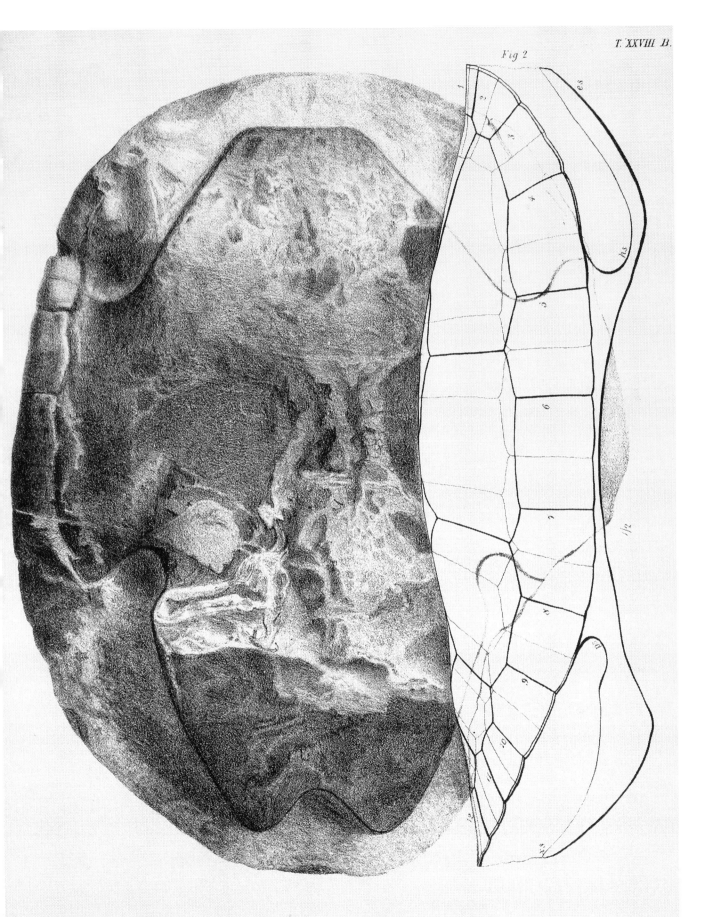

Fig 2

MONOGRAPH

ON

THE FOSSIL REPTILIA

OF THE

LONDON CLAY, AND OF THE BRACKLESHAM AND OTHER TERTIARY BEDS.

PART II.   SUPPLEMENT No. 1.

Pages 1—4; Plate XXIX.

CHELONIA (Platemys).

BY

PROFESSOR OWEN, D.C.L., F.R.S., F.L.S., F.G.S., &c.

Issued in the Volume for the Year 1849.

LONDON:

PRINTED FOR THE PALÆONTOGRAPHICAL SOCIETY.

1850.

MONOGRAPH

ON THE

FOSSIL REPTILIA OF THE LONDON CLAY.

SUPPLEMENT TO THE ORDER—*CHELONIA.*

Family—PALUDINOSA.

PLATEMYS BOWERBANKII (?).   Tab. XXIX, figs. 1, 2.

The evidence of species of Chelonia of the Fresh-water or Marsh-dwelling family, *Paludinosa*, has hitherto been derived only from such parts of the skeleton of the trunk as have been described, figured, and referred to the genera *Platemys* and *Emys*, in Part I of the present Monograph, pp. 62-76.

Since those pages were sent to press, Mr. Bowerbank has been so fortunate as to obtain from the Eocene clay at Sheppy the portion of fossil skull, of which two views are given of the natural size in T. XXIX, figs. 1, 2.   If these figures, and especially the side view, fig. 1, be compared with the corresponding view of the skulls, T. I, fig. 1; T. III, fig. 1; T. XV, fig. 1, or T. XI, fig. 2, a marked difference will be discerned in the form and proportion of the orbit, which is smaller and more nearly circular in fig. 1, T. XXIX.

But the bony chamber for the eyeball forms one of the characters by which the skull is distinguished in the marine and fresh-water families of the order *Chelonia.* The orbit, for example, is always much larger in proportion to the entire skull in the marine species, and commonly of the oval form, which is preserved in the beautiful fossil skull of the *Chelone cuneiceps*, T. XV, fig. 1; or with the upper and outer part even more produced and angular than is there represented.   In the families *Fluvialia* (Trionyx) and *Paludinosa* (Emydians), the orbit is not merely much smaller in proportion to the skull, it is circular, or nearly so, and not produced at the upper and outer angle.   By this character, we are led to refer the fossil skull under description to the fresh-water division of the Chelonian order.

Our choice between the Fluviatile or Paludinose families of that division is guided by the formation of the border of the orbit, and by the proportionate length and the form of the face or muzzle in advance of it.

1

In the species of *Emys* (*Podocnemys expansa*) which I have selected for comparison, as offering upon the whole the nearest approach, which any Chelonian skull at my command gives, to the unique fossil in question, the malar bone (*i*, in Cuvier's figure of the skull of *Emys expansa*, pl. xi, fig. 9, of the 'Ossemens Fossiles,' tom. v, pt. ii, 1825; 26 in the figure of the fossil, fig. 1, T. XXIX), becomes much contracted as it approaches the orbit, to which it contributes a small part of the posterior border. In the *Chelones* the malar bone forms a larger proportion of the orbital rim (see Cuvier, tom. cit., pl. xi, fig. 1 *i*), and contributes more to its under than its back part, which is chiefly formed by the characteristically large postfrontal 12 (*g* in Cuvier's figs.); and this character was manifested in the ancient Eocene turtles as well as in the modern species, as may be seen by reference to the bones numbered 26 and 12, in T. I, fig. 1; T. XV, fig. 1, of the present work. The superior maxillary bone 21 (*b* in Cuvier's figs.) is longer in the *Emys*, extends further back in the orbit, and is deeper at its posterior termination, than in the *Chelones*. In all these characters, derived from the bones entering into the formation of the orbit, the fossil under comparison agrees with the *Emys*, and, indeed, departs further from the *Chelones* than the *Podocnemys expansa* does, by the much smaller proportion in which the anteriorly contracted malar bone (26) contributes to the rim of the orbit. In the *Trionyces*, the malar bone forms a larger proportion of the border of the orbit than in the *Podocnemys expansa*, and *a fortiori*, than in the fossil in question.

The choice between the Fluviatile or Paludinose tribes of the fresh-water Chelonians, in the determination of this fossil, is better guided by the form and proportions of the skull anterior to the orbit. In the recent *Trionyces* the muzzle is more acute, and in most of them more prolonged than in the Emydians, with which the fossil skull agrees in the shortness of the muzzle; whilst it departs further than most recent Emydians from the *Trionycidæ*, in the broad truncated character of its anterior termination. There is also a very well-marked character of affinity to the *Podocnemys expansa*, in the smooth and shallow canal which extends from the fore part of the orbit forwards to the border of the external nostril across the upper part or nasal process of the superior maxillary bone (21). This groove is very accurately represented in fig. 1, T. XXIX, in the fossil; it is rather broader in proportion to its length in the *Podocnemys expansa*; but so far as it has depended upon the presence and arrangement of the facial scutes, it is decisive against the fossil having appertained to any species of soft turtle (*Trionyx*), in which such epidermal parts were entirely wanting.

The marks of the supracranial scutes in the fossil are, as in some Emydians, too feebly and obscurely traceable to permit of a satisfactory comparison of their arrangement. The exterior surface of the prefrontal (16), frontal (11), postfrontal (12), and parietal (7) bones is subreticulate. The substance of the bones is thick and coarsely cancellous. The nasal bone is connate with the prefrontal, as in most modern Emydians; in the proportion of this compound bone the fossil resembles more the ordinary

Emydians (*Emys europæa*, e. g.) than it does the *Podocnemys expansa*. The border of the prefronto-nasal bone forming the upper part of the nostril is thick and rounded; as is also the lateral border of the same cavity formed by the maxillary. The lower part of this border of the maxillary shows the suture for the premaxillary, which must have presented similar proportions to the premaxillary of the *Podocnemys expansa* and other Emydians. The shape of the frontal (11), the proportion of the upper border of the orbit which it forms, and the course of its sutures with the contiguous bones, are clearly indicated in fig. 2, T. XXIX. The straight line formed by the suture between the frontal (11) and postfrontal (12) resembles that in the *Podocnemys expansa*; it is bent or curved in the *Chelones*. To what extent the postfrontal (12) was continued backwards, whether so, as with the parietal, to roof over the temporal fossa, as in the *Podocnemys expansa*,[*] or, in a less degree, leaving that fossa open superiorly, as in the Emydians generally, is a question which will require for its determination a more perfect specimen than the fossil under description. The thickness, however, of the fractured posterior part of the postfrontal indicates that the bone had been broken not very close to its natural posterior border, on the supposition that this was free, as in the Emydians generally; and the part of the suture of the postfrontal with the parietal which has been preserved, extends obliquely outwards and backwards, as in *Podocnemys expansa*, not directly backwards, as in most of the *Emydes* with open temporal fossæ. (Compare Cuvier, loc. cit., fig. 10 with fig. 14, the suture between *g* and *h*.) With respect to the parietal bones (7̄), these are too much mutilated to show more than the position and extent of the coronal suture.

A few words may be perhaps expected relative to the difference which the fossil in question presents to the land-tortoises. In comparison with the skull of a *Testudo indica* of corresponding dimensions with the fossil, the larger proportional size of the orbits distinguishes the skull of that terrestrial species almost as strongly as the same character does the skull of the marine turtles. But in addition to this, the malar bone forms a larger proportion of the back part of the orbit in the *Testudo*, and the prefronto-nasal part of the skull is more bent down; the suture between the frontals and prefrontals describes a curve convex forwards in the *Testudo*, whilst it deviates very little from a straight line in the fossil, and that little is convex backwards. The extent also of the upper surface of the postfrontals and parietals, so far as these are preserved in the fossil, is greater than the whole of those bones in the land-tortoise compared.

Having been led by the foregoing comparisons to refer the fragment of the fossil skull (T. XXIX, figs. 1, 2) to the family *Paludinosa*, it is reasonable to conjecture that it may have appertained to some one of the large Emydians, which we already know to have left their carapaces in the Eocene clay of Sheppy. One commonly finds in

---

[*] See Cuvier, loc. cit., pl. xi, fig. 1 *a*.

the recent skeletons of Emydians, that any particular character of the exterior surface of the bones of the trunk is repeated on the upper surface at least of the bones of the head. This comparison, in the present instance, indisposes me to regard the fossil in question as having belonged to the *Emys levis,* or to the *Emys bicarinata,* or to the *Platemys Bullockii* with the punctate plastron. I should be rather led to select the *Platemys Bowerbankii* from the character in question, as exhibited by the carapace and plastron described at p. 66. But, in provisionally registering the fossil skull in question under the name of *Platemys Bowerbankii,* I should wish to be understood as by no means vouching for the accuracy of the reference. The conjecture rests solely on the character above referred to, which is far from being decisive; and its only value is, that it happens to be the only one by which we can be guided at present in forming any opinion at all as to the specific relations of the fossil in question.

TAB. XXIX.

Chelonia.

Fig.
1. Side view of the fore part of the skull of the *Platemys Bowerbankii,* nat. size.
2. Upper view of the same fossil.
3. Side view of a fractured tympanic bone of a large Turtle (*Chelone*), from Bracklesham, showing the long and slender ossicle or ' columella' (16) *in situ ;* nat. size.
4. Extremity of the same tympanic bone, to which the ' membrana tympani' was attached.
5. Proximal end of the femur of a very large Chelonian from the Isle of Sheppy, nat. size.
5'. Left femur of a Turtle (*Chelone mydas*) which weighed 150 lbs., nat. size.

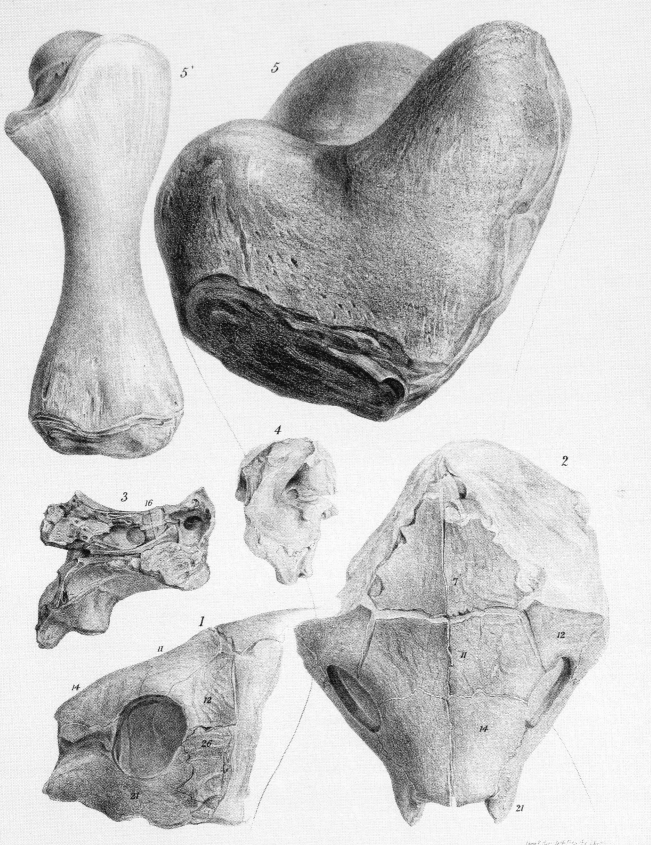

MONOGRAPH

ON

THE FOSSIL REPTILIA

OF THE

LONDON CLAY, AND OF THE BRACKLESHAM AND OTHER TERTIARY BEDS.

PART II.

Pages 5—50; Plates I—XI, and IIa.

CROCODILIA (Crocodilus, &c.).

BY

PROFESSOR OWEN, D.C.L., F.R.S., F.L.S., F.G.S., &c.

sued in the Volume for the Year 1849.

LONDON:

PRINTED FOR THE PALÆONTOGRAPHICAL SOCIETY.

1850.

PART II.

Order—*CROCODILIA.*

CROCODILES, ALLIGATORS, GAVIALS.

OF the numerous and various kinds of Reptiles, the fossil remains of which have been discovered in the tertiary and secondary strata of Great Britain, many are found to have their nearest representatives, amongst the actual members of the class, in the present order; and here more particularly in the long and narrow-snouted genus called, through a corrupt latinization of its native name, *Gavialis,* which is now represented by the Gavial or, more properly, Garrhiāl, of the river Ganges.

In the interpretation of the fossil remains of Reptiles, no skeleton has more frequently to be referred to than that of the Gavial or Crocodile, or has thrown more light on the nature of those singularly modified forms of the class which have long since passed away.

It is accordingly requisite for the palæontologist who would describe the fossil remains of reptiles, to make himself, in the first place, thoroughly conversant with the osteology of the recent *Crocodilia.* This knowledge can be gained only by assiduous study of the skeletons themselves, with the aid of the best descriptions, or the guide of a competent teacher. But to enable the reader to follow or comprehend the description of the fossil Saurians, some elementary account of the Crocodilian skeleton is at least necessary, accompanied with illustrations of the parts which, in the sequel, will have to be frequently referred to under special or technical names.

In Tab. XI of the present part of this Monogragh is given a reduced or miniature side view of the skeleton of a Gavial which was twenty-five feet in length—dimensions which are rarely found to be surpassed in the present day. Beneath it is a restoration of the skeleton of the Teleosaur, or extinct Gavial of the Triassic or Oolitic period, showing how closely the general type of conformation has been adhered to, the modifications of the more ancient form of Crocodile evidently adapting it for moving with greater speed and facility through the water, and indicating it to have been more strictly aquatic, and probably marine.

The particular nature of these modifications will be explained when I come to describe the Crocodiles of the secondary strata. I propose at present to give a preliminary sketch of the osteology of the recent *Crocodilia.*

A glance at a natural or well-articulated skeleton of one of these reptiles, such as is figured in T. XI, will show that it consists mainly of a series of segments, more or less alike. From the back of the head to the end of the tail, the chief part of each segment consists of a cylindrical portion or 'body,' differing only in its proportions, and diminishing as it recedes from the trunk. Every segment sends a plate of bone upwards from its upper or dorsal surface, which plate or 'spine' is supported by an arch of bone, except in the diminishing segments at the end of the tail.

Other plates of bone, of more variable forms and dimensions, project from each side of the segments of the trunk and basal part of the tail. In a less proportion, but still in a great number of the segments, an arch of bone is formed below, or on the ventral side of the cylindrical body; but this lower arch is more variable in its proportions and mode of composition than the upper arch: it is open or incomplete in the neck. Under all these variations, however, there is plainly manifested a fundamental unity of plan in the composition of the different segments, which have accordingly received the common appellation of 'vertebra.'

For the convenience of description, the vertebræ are divided, though somewhat arbitrarily, into groups bearing special or specific names. Those next the head, with the inferior arch incomplete below, are called 'cervical vertebræ;' they are usually nine in number: those that follow with the inferior arch closed below, or which have the laterally projecting parts slender and freely moveable, are called 'dorsal vertebræ;' the other vertebræ of the trunk that have no lateral moveable appendages, are called 'lumbar vertebræ;' the last vertebræ of the trunk, always two in number in the *Crocodilia*, the inferior arches of which coalesce to support and be supported by the hind limbs, are the 'sacral vertebræ;' the segments of the tail are the 'coccygeal,' or 'caudal vertebræ,' whether they possess or not an inferior arch, or whatever other modifications they may offer.

These names, 'cervical,' 'dorsal,' 'lumbar,' 'sacral,' 'coccygeal,' were originally applied to corresponding segments or vertebræ in the human skeleton, from the study of which the nomenclature of osteology takes its date: it may well be supposed, therefore, that a classification and designation of vertebræ based upon knowledge limited to their characters in a single example of the vertebrated series, and that example one in which the common type has been most departed from, to adapt it to the peculiar attitude and powers of the human species, would fall far short of what is required to express the general ideas derived from a comparison of all the leading modifications of the vertebrate skeleton; and accordingly the anatomist who passes from a previous acquaintance with human osteology only, to the study of those of the lower Vertebrata, finds that he has to rectify, in the first place, the erroneous notions which anthropotomy has taught him of the nature of the primary segment of his own and other vertebrated skeletons, and to acquire true ideas, with the concomitant nomenclature, of the essential constituents or anatomical elements of such segment.

In human anatomy, for example, the costal elements are only recognised when they retain throughout life that distinctness, or moveable union with the rest of their segment, which they manifest at their first appearance; and they are then classified as distinct bones from the rest of their segment, to which the term 'vertebra' is restricted, and which is equally regarded as a single bone; as, e. g., in the dorsal region of the skeleton. In the cervical region the whole segment is called 'vertebra,' and is recognised as the equivalent bone to a dorsal vertebra, although it includes the costal elements, because these have coalesced with the rest of their segment, which anchylosis is misinterpreted as a mere modification of a transverse process; and the 'cervical vertebra' is distinguished by having that process 'perforated,' and not entire as in the other vertebræ.

But, in the Crocodile, the embryonic condition of the cervical ribs in Man is retained throughout life; and, therefore, if we were to be guided by the characters laid down by the recognised authorities in anthropotomy for the classification of its vertebræ, we should seek in vain for any vertebræ with " transverse processes perforated for the transmission of vertebral arteries," whilst we should find all the vertebræ from the head to the loins, "with articular surfaces, either on their sides or their transverse processes, where they join with ribs," and should accordingly have to reckon these as " dorsal vertebræ."

These and many similar instances which might be adduced, have compelled me to premise a few brief explanations of the principles and nomenclature by which I shall describe the fossil remains of the *Reptilia*, and illustrate their nature by reference to the skeletons of their existing representatives, in the present and succeeding Monographs.

The primary segment of the skeleton of all Vertebrata is a natural group of bones, which may be severally recognised and defined under all the modifications to which such segment may have been subjected in subservient adaptation to the habits and exigencies of a particular species.

A view of such a segment, as it exists in the thorax of the crocodile, the tortoise, and the bird, is given at p. 5, Part I, of the present Monograph.

The part marked c is the 'centrum,' or body of the vertebral segment; it is always developed originally as a separate element, and retains its character of individuality in the tortoise and crocodile. The bony arch above the centrum was formed originally by two distinct side-plates,—the 'neurapophyses,' n, which coalesce with one another at their summits and thence develope a median plate or process of bone called the 'neural spine' ns. Other bony processes which shoot out from the neurapophyses are more variable, and will be afterwards noticed. The arch so formed coalesces with the centrum in the bird, and constitutes an apparently single bone, to which, in anthropotomy, the name 'vertebra' would be restricted. But it would be as reasonable to confine it to the central element (c) in the tortoise and crocodile; for the parts of the inferior arch are not less essentially parts of the same natural segment, than the neurapophyses which have formed the upper arch. The next pair of elements, then, which we have to notice, is marked in figs. 4, 5, and 6, *pl*, signifying 'pleura-pophysis,' the name of these elements. In the segments figured they retain their primitive distinctness, and acquire unusual length, in order to aid in encompassing the dilated canal or cavity for the heart and lungs: so modified, these elements are commonly called 'ribs,' or 'vertebral ribs.'

The elements more constantly employed to protect the vascular or 'hæmal' axis, in other words, to form the inferior or hæmal canal, are those marked *h* in figs. 4 and 6: they are the 'hæmapophyses,' which are usually articulated, like the neurapophyses, with the centrum, but are displaced by the great centres of the vascular system in the thorax, where they have got the special name of 'sternal ribs,' and also that of 'costal cartilages,' or 'cartilages of the ribs,' when they do not become ossified. The hæmal arch in the thorax is usually completed by a median element (*hs*), called a 'hæmal spine,' but which itself becomes vastly expanded in the bird (fig. 4); it is, nevertheless, the part in the hæmal arch which repeats below, or answers to the part (*ns*) in the upper arch. In the segments of the trunk and tail, the element (*ns*) retains its normal size and form as a 'neural spine;' but where the central axis of the nervous system becomes unusually developed, as in the head, e. g., analogously to the development of the vascular centres in the chest, the neural canal is correspondingly expanded, and the cavity acquires a special name, and is called 'cranium,' just as the analogously expanded hæmal canal is called 'thorax.' Into the formation of the wall of the cranium other vertebral elements enter besides the neurapophyses, those e. g. which are numbered 8 and 12 in fig. 9, p. 17, of the present Part; the neural spine (7 and 11 in the same figure) retains its primitive distinctness, is expanded horizontally, and, like the 'sternum' in the thorax of the bird (*hs*, fig. 4, p. 5, Part I), it receives a special name (7), e. g. of 'parietal', and (9) of 'frontal' in fig. 9. The elements *a a* (figs. 4 and 6, Part I) form a symmetrical pair of bones or cartilages, attached at one end to the hæmal arch, and projecting out-wards and backwards. These are the 'prosartemata,' or appendages; they are, of all the elements of the vertebral segment, those that are least constant in regard to their presence, and, when present, are subject to the greatest amount of development and metamorphosis: they become, e. g., the pterygoid appendages in the nasal segment of the fish's skull, the opercular bones in the frontal segment of the fish, the branchiostegal rays in the parietal segment, the pectoral fins in the occipital segment, and they are deve-loped into the fore limbs and hind limbs, the arms, wings, and legs of other Vertebrata.*

As the nervous and vascular centres become reduced in size, the bony canals or arches protecting them are simplified and contracted, and the vertebra assumes a symmetrical character. In the Crocodile, the hæmal arch, in the tail, e. g., is formed by the hæmapophyses, which ascend and articulate directly with the centrum; the pleurapophyses are shortened, directed outwards, and become anchylosed to

---

* The facts and arguments in support of this conclusion, are detailed in my works 'On the Nature of Limbs,' and 'On the Archetype of the Vertebrate Skeleton,' 8vo (Van Voorst).

form 'transverse processes;' but such a vertebra, when analysed as it is developed, resolves itself very nearly into the ideal type given in the subjoined diagrammatic cut (fig. 7); n is the neural axis, called 'myelon,' or 'spinal marrow;' h is the hæmal axis, the chief trunk of which is called 'aorta,' and 'caudal artery.' The names of the vertebral elements which, being usually developed from distinct centres, are called 'autogenous,' are printed in Roman type; the *Italics* denote the 'exogenous' parts, more properly called 'processes,' which shoot out from the preceding elements.

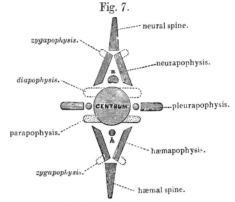

Fig. 7.

Ideal typical vertebra.

On comparing this form of the primary segment with that figured in Cut 4, p. 5, Part I, it will be seen that they differ by altered proportions with some change of position of certain elements; but every modification resulting in the various forms of the parts of the skeleton figured in T. XI, has its seat in one or other of the segmental or 'vertebral' elements above defined; and the same principle I believe that I have established with regard to the internal skeleton in all vertebrate animals.

With this preliminary explanation, the nature and relations to the typical vertebra of the parts of the Crocodilian vertebræ, figured in T. IV, V, IX, will be, it is hoped, readily appreciated. In T. IX, in which are figured, through opportunities kindly afforded by the Marchioness of Hastings, and Searles Wood, Esq., F.G.S., some of the most perfectly-preserved fossil reptilian vertebræ which have hitherto been discovered, the elements and processes are indicated by the initial letter of their names. Figs. 1 and 2 give a side view and a back view of a cervical vertebra, apparently the fourth, of the *Crocodilus Hastingsiæ*, from the Eocene deposits at Hordwell; c is the centrum, n the neural canal formed by the neurapophyses, which have coalesced superiorly with each other, and with the neural spine (ns). Inferiorly they articulate by a suture (which is shown by the wavy line on each side of the process d in fig. 1) with the centrum; pl is the pleurapophysis, which articulates by two parts, the lower one called the 'head to the process from the centrum, the upper one called the 'tubercle' to the process from the neurapophysis; beyond the union of the head and tubercle, the pleurapophysis projects freely outwards and downwards, but instead of being elongated in that direction, it becomes expanded in the direction of the axis of the body, i. e. forwards and backwards, and so acquires a shape which has given rise to the name 'hatchet bone' or 'hatchet-shaped process,' * applied to this element in the *Plesiosaurus*.

* "To compensate for the weakness that would have attended this great elongation of the neck, the *Plesiosaurus* had an addition of a series of hatchet-shaped processes on each side of the lower part of the cervical vertebra." (Buckland, Bridgewater Treatise, vol. i, p. 206, and vol. ii, p. 30, 1836.)

Cuvier recognised in these lateral bones, "en forme de hache," the homologues of the "petites côtes cervicales" of the Crocodile. (Ossemens Fossiles, 4to, tom. v, pt. ii, p. 479, 1824.) And Conybeare had

2

The purport of this modification is the same in the *Crocodilia* as that which seems to be more called for in the *Plesiosaurus*, viz. to augment the strength of the cervical region of the skeleton ; and this is so effectually done by the overlapping of the hatchet-shaped ribs of this region in the *Crocodilia*, as shown in T. XI, that the flexibility of the neck is much restricted, although the joint of the head allows that part to be bent from side to side at nearly right angles with the neck.   When, however, the head is held firmly forwards by its powerful muscles, the imbricated vertebræ of the neck transmit with great effect the impulse which the strong and long tail gives to the rest of the body in the act of swimming.

In T. IX, fig. 3 the cervical vertebra is represented minus its pleurapophyses, and it answers accordingly to that portion of the natural segment to which the term ' vertebra' is usually restricted in the dorsal region of the trunk.   The exogenous processes shown in this view of the vertebra are, *p*, the ' parapophysis' or inferior transverse process, developed from the centrum ; *d*, the ' diapophysis' or upper transverse process developed, as in most cases it is, from the neurapophysis; *z*, *z'*, are the ' zygapophyses' or ' oblique processes,' which, from their function in articulating together contiguous vertebræ, are also called ' articular processes.'   In most of the cervical, and in some of the dorsal, vertebræ of the Crocodile, an exogenous process is developed from the under surface of the centrum, called ' hypapophysis ;' it is indicated by the letters *hy* in fig. 2, T. IX.   In some species it is double,* and beneath the atlas it becomes ' autogenous' or is developed as a separate element, *ca, ex*, of the subjoined Cut, fig. 8, in which condition the part is found beneath the centrums of two or three of the anterior cervical vertebræ in the Ichthyosaurus.†

The first and second vertebræ of the neck are peculiarly modified in most air-breath-

Fig. 8.

ing Vertebrata, and have accordingly received the special names, the one of ' atlas,' the other of ' epistropheus' or ' axis.'   In Comparative Anatomy these become arbitrary terms, the properties being soon lost which suggested those names to the human anatomist; the ' atlas' e. g. has no power of rotation upon the ' axis' in the Crocodile, and it is only in the upright skeleton of man that the large globular head is sustained upon the shoulder-like processes

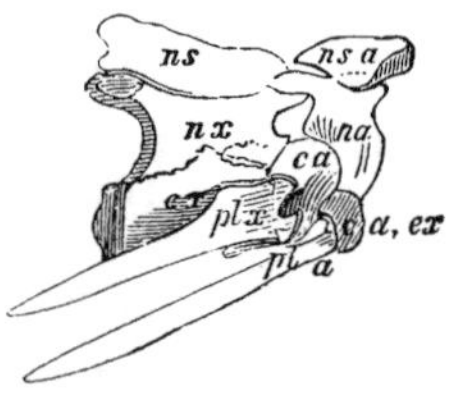

Atlas and Axis vertebræ of the Crocodile.

of the ' atlas.'   In the Crocodile, these vertebræ are concealed by the peculiarly prolonged angle of the lower jaw in the side view of the skeleton in T. XI,

previously extended the same homology to the "particularly prominent wing-like appendages to the transverse processes in many of the long-necked quadrupeds, and the long styloid processes of the cervical vertebra of birds." (See his admirable Memoir of June 14th, 1822, in the Geol. Trans., 2d series, vol. i, p. 384.)

* In *Crocodilus basifissus*, e. g., see the Quarterly Journal of the Geological Society, November 1849, p. 381, pl. x, fig. 2.

† This interesting discovery was communicated by its author, Sir Philip de M. Grey Egerton, Bart., F.G.S., to the Geological Society of London, in 1836, and is published in the fifth volume of the second series of their Transactions, p. 187, pl. 14.

and a separate view of them is, therefore, given in figure 8. The pleurapophyses are retained in both segments, as in all the other vertebræ of the trunk. That of the atlas, fig. 8, *pl a*, is a simple slender style, articulated by the head only, to the independently developed inferior part of the centrum, or 'hypapophysis' (*ca, ex*). The neurapophyses (*na*) of the atlas retain their primitive distinctness; each rests in part upon the proper body of the atlas (*ca*), in part upon the hypapophysis. The neural spine (*ns, a*) is also here an independent part, and rests upon the upper extremities of the neurapophyses. It is broad and flat, and prepares us for the further metamorphosis of the corresponding element in the cranial vertebræ.

The centrum of the atlas (*ca*), called the odontoid process of the epistropheus in Human Anatomy, here supports the abnormally-advanced rib of that vertebra, which in some *Crocodilia* is articulated by a bifurcate extremity, like the ribs of the succeeding cervical vertebræ; but it is not expanded or hatchet-shaped at the free extremity. The proper centrum of the axis vertebra (*cx*) is the only one in the cervical series which does not support a rib; it articulates by suture with its neurapophyses (*nx*), and is characterised by having its anterior surface flat, and its posterior one convex.

With the exception of the two sacral vertebræ, the bodies of which have one articular surface flat and the other concave, and of the first caudal vertebra, the body of which has both articular surfaces convex, the bodies of all the vertebræ beyond the axis have the anterior articular surface concave, and the posterior one convex, and articulate with one another by ball-and-socket joints. This type of vertebra, which I have termed 'proccœlian,'* characterises all the existing genera and species of the family *Crocodilia*, with all the extinct species of the tertiary periods, and also two extinct species of the Greensand formation in New Jersey.† Here, so far as our present knowledge extends, the type was lost, and other dispositions of the articular surfaces of the centrum occur in the vertebræ of the *Crocodilia* of the older secondary formations. The only known Crocodilian genus of the periods antecedent to the Chalk and Greensand deposits with vertebræ articulated together by ball-and-socket joints, have the position of the cup and the ball the reverse of that in the modern Crocodiles, and the genus, thus characterised by vertebræ of the 'opisthocœlian' type, has accordingly been termed *Streptospondylus*, signifying 'vertebræ reversed.' The aspects of the zygapophyses are, however, more constant; the anterior ones, T. IX, fig. 3 *z*, look obliquely inwards; the posterior ones, ib. *z'*, obliquely outwards. In looking, therefore, upon the cut surface of a vertical longitudinal section of a Crocodilian vertebra, the smooth, flattened inner surface of the anterior zygapophysis is turned towards the observer, and the convex outer surface of the posterior zygapophysis. Thus the anterior and posterior extremity of the vertebra being determined by observation of the aspect and direction of the zygapophyses, it is at once seen whether the body has the

---

* Προς, before ; κοιλος, concave.
† Quarterly Journal of the Geological Society, November 1849.

procœlian structure, as in the true Crocodiles, T. IV, T. IX, or the opisthocœlian structure, as in the *Streptospondylus*. But the most prevalent type of vertebra amongst the Crocodilia of the secondary periods was that in which both articular surfaces of the centrum were concave, but in a less degree than in the single concave surface of the vertebræ united by ball and socket. A section of a vertebra of this 'amphicœlian' type, such as existed in the *Teleosaurus* and *Steneosaurus*, will be figured in a subsequent Monograph. In the *Ichthyosaurus*, the concave surfaces are usually remarkable for their depth, the vertebræ resembling in this respect those of fishes. Some of the most gigantic of the *Crocodilia* of the secondary strata had one end of the vertebral centrum flattened, and the other (hinder) end concave; this 'platycœlian' type we find in the dorsal and caudal vertebræ of the gigantic *Cetiosaurus*.

With a few exceptions, all the modern Reptiles of the order *Lacertilia* have the same procœlian type of vertebræ as the modern *Crocodilia*, and the same structure prevailed as far back as the period of the *Mosasaurus*, and in some smaller members of the Lacertilian order in the Cretaceous and Wealden epochs.

Resuming the special description of the osteology of the modern *Crocodilia*, we find the procœlian type of centrum established in the third cervical, which is shorter but broader than the second; a parapophysis is developed from the side of the centrum, and a diapophysis from the base of the neural arch; the pleurapophysis is shorter, its fixed extremity is bifid, articulating to the two above-named processes; its free extremity expands, and its anterior angle is directed forwards to abut against the inner surface of the extremity of the rib of both the axis and atlas, whilst its posterior prolongation overlaps the rib of the fourth vertebra.

The same general characters and imbricated coadaptation of the ribs characterise the succeeding cervical vertebræ to the seventh inclusive, the hypapophysis (*hy*, fig. 2, T. IX) progressively though slightly increasing in size. In the eighth cervical the rib becomes elongated and slender; the anterior angle is almost or quite suppressed, and the posterior one more developed and produced more downwards, so as to form the body of the rib, which terminates, however, in a free point. In the ninth cervical the rib is increased in length, but is still what would be termed a 'false' or 'floating rib' in anthropotomy.

In the succeeding vertebra the pleurapophysis articulates with a hæmapophysis, and the hæmal arch is completed by a hæmal spine; and by this completion of the typical segment we distinguish the commencement of the series of dorsal vertebræ. With regard to the so-called 'perforation of the transverse process,' this equally exists in the present vertebra, as in the cervicals, as may be seen by comparing fig. 6, p. 5, Part I of this Monograph, with fig. 2, T. IX, Part II; in both, the foramen is the vacuity intercepted between the bifurcate extremity of the rib and the rest of the vertebra with which that rib articulates; and, on the other hand, the cervical vertebræ equally show surfaces for the articulation of ribs. Cuvier, in including the proximal portions of the ribs with the rest of the vertebra, in his figure of a dorsal vertebra of a

Crocodile,* so far follows nature, and produces a parallel to his figure of a cervical vertebra; but the entire natural vertebra or segment includes the parts delineated in outline in Cut 6, p. 5, Part I. In that figure is shown the semiossified bar h' which is interposed between the pleurapophysis pl and hæmapophysis h in the Crocodilia and some existing Lizards. The typical characters of the segment due to the completion of both neural and hæmal arches, is continued in some species of *Crocodilia* to the sixteenth, in some (*Crocodilus acutus*) to the eighteenth vertebra. In the *Crocodilus acutus* and the *Alligator lucius*, the hæmapophysis of the eighth dorsal rib (seventeenth segment from the head) joins that of the antecedent vertebra. The pleurapophyses project freely outwards, and become 'floating ribs' in the eighteenth, nineteenth, and twentieth vertebræ, in which they become rapidly shorter, and in the last appear as mere appendages to the end of the long and broad diapophyses: but the hæmapophyses by no means disappear after the solution of their union with their pleurapophyses; they are essentially independent elements of the segment, and they are continued, therefore, in pairs along the ventral surface of the abdomen of the Crocodilia, as far as their modified homotypes the pubic bones. They are more or less ossified, and are generally divided into two or three pieces.

Another character afforded by the hæmal arch is the more important in reference to palæontology, as it affects the centrum and neural arch of the vertebra as well as the pleurapophysis; and thus aids in the determination of the vertebra. The parapophysis progressively ascends upon the side of the centrum in the two anterior dorsal vertebræ, and disappears in the third, or, passing upon its neurapophysis, blends with the base of the diapophysis. In this segment, therefore, the proximal end of the rib ceases to be bifurcate, but is simply notched, the curtailed head being applied to the end of the thickened anterior part of the transverse process, and the tubercle abutting against its extremity; in the five following dorsals the head and tubercle of the rib progressively approximate and blend together, or the head disappears in the tenth dorsal, in which the rib is simply attached to the end of the diapophysis. The hypapophysis ceases to be developed after the third or fourth dorsal vertebræ. The zygapophyses become gradually more horizontal, the anterior ones looking more directly upwards, the posterior ones downwards.

The 'lumbar vertebræ' are those in which the diapophyses cease to support moveable pleurapophyses, although they are elongated by the coalesced rudiments of such which are distinct in the young *Crocodilia*. The development and persistent individuality of more or fewer of these rudimental ribs determines the number of the dorsal and lumbar vertebræ respectively, and exemplifies the purely artificial character of the distinction. The number of vertebræ or segments between the skull and the sacrum, in all the *Crocodilia* I have yet examined, is twenty-four. In the skeleton of

* Ossemens Fossiles, 4to, tom. v, pt. ii, pl. iv, fig. 4.

a Gavial I have seen thirteen dorsal and two lumbar; in that of a *Crocodilus cata-phractus* twelve dorsal and three lumbar; in those of a *Crocodilus acutus*, and *Alligator lucius*, eleven dorsal and four lumbar, and this is the most common number; but in the skeleton of the Crocodile, I believe of the species called *Croc. biporcatus*, described by Cuvier,[*] he gives five as the number of the lumbar vertebræ. But these varieties in the development or coalescence of the stunted pleurapophysis are of little essential moment; and only serve to show the artificial character of the 'dorsal' and 'lumbar' vertebræ. The coalescence of the rib with the diapophysis obliterates of course the character of the 'costal articular surfaces;' which we have seen to be common to both dorsal and cervical vertebræ. The lumbar zygapophyses have their articular surfaces almost horizontal, and the diapophyses, if not longer, have their antero-posterior extent somewhat increased; they are much depressed, or flattened horizontally.

The sacral vertebræ are very distinctly marked by the flatness of the coadapted ends of their centrums: there are never more than two such vertebræ in the *Crocodilia* recent or extinct; in the first the anterior surface of the centrum is concave, in the second it is the posterior surface; the zygapophyses are not obliterated in either of these sacral vertebræ, so that the aspects of their articular surface—upwards in the anterior pair, downwards in the posterior pair—determines at once the corresponding extremity of a detached sacral vertebra. The thick and strong transverse processes form another characteristic of these vertebræ; for a long period the suture near their base remains to show how large a proportion is formed by the pleurapophysis. This element articulates more with the centrum than with the diapophysis developed from the neural arch;[†] it terminates by a rough, truncate, expanded extremity, which almost or quite joins that of the similarly but more expanded rib of the other sacral vertebræ. Against these extremities is applied a supplementary costal piece, serially homologous with the superadded piece to the proper pleurapophysis in the dorsal vertebræ (l', fig. 6, p. 5, Part I), but here interposing itself between the pleurapophyses and hæmapophyses of both sacral vertebræ, not of one only. This intermediate pleurapophysial piece is called the 'ilium;' it is short, thick, very broad, and subtriangular, the lower truncated apex forming with the connected extremities of the hæmapophysis an articular cavity for the diverging appendage, called the 'hind leg.' The hæmapophysis of the anterior sacral vertebra is called 'pubis;' it is moderately long and slender, but expanded and flattened at its lower extremity, which is directed forwards towards that of its fellow, and joined to it through the intermedium of a broad, cartilaginous, hæmal spine, completing the hæmal canal. The posterior hæmapophysis is broader, subdepressed, and subtriangular, expanding as it approaches its fellow to complete the second hæmal

---

[*] Tom. cit., p. 95. It is to be observed that Cuvier begins to count the dorsal vertebra when the rib has changed its hatchet-shape for a styloid shape.

[†] Cuvier, who well describes this structure, remarks, "aussi méritent-elles plutôt le nom des côtes que celui d'apophyses transverses." (Tom. cit., p. 98.)

arch; it is termed 'ischium.' The great development of all the elements of these hæmal arches, and the peculiar and distinctive forms of those that have thereby acquired, from the earliest dawn of anatomical science, special names, relates physiologically to the functions of the diverging appendage which is developed into a potent locomotive member. This limb appertains properly, as the proportion contributed by the ischium to the articular socket and the greater breadth of the pleurapophysis show, to the second sacral vertebra; to which the ilium chiefly belongs.

The first caudal vertebra, which presents a ball for articulating with a cup on the back part of the last sacral, retains, nevertheless, the typical position of the ball on the back part of the centrum; it is thus biconvex, and the only vertebra of the series which presents that structure. I have had this vertebra in three different species of extinct Eocene *Crocodilia*. In the *Crocodilus toliapicus*, T. IV, fig. 7; in the *Croc. champsoïdes*, T. V, fig. 10; and in the *Crocodilus Hastingsiæ*, T. IX, fig. 7.

The advantage of possessing such definite characters for a particular vertebra is, that the homologous vertebra may be compared in different species, and may yield such distinctive characters as will be hereafter pointed out in those of the three species above cited.

The first caudal vertebra, moreover, is distinguished from the rest by having no articular surfaces for the hæmapophyses, which in the succeeding caudals form a hæmal arch, like the neurapophyses above, by articulating directly with the centrum. The arch so formed has its base not applied over the middle of a single centrum, but like the neural arch in the back of the tortoise and sacrum of the bird, across the interspace between two centrums. The first hæmal arch of the tail belongs, however, to the second caudal vertebra, but it is displaced a little backwards from its typical position.

The detached centrum of a caudal vertebra, besides being more slender and compressed, is distinguished from those of the before-described vertebræ by the two articular surfaces at the posterior border of their under surface T. IV, fig. 9. The zygapophyses become vertical as far as the sixteenth or seventeenth, beyond which the two posterior zygapophyses coalesce in an oblique plane notched in the middle, which is received into a wider notch at the fore part of the neural arch of the succeeding vertebra. The sutures between the pleurapophyses and diapophyses are maintained during a long period of the animal's growth, and demonstrate the share which these two elements respectively take in the formation of the transverse process. So constituted, these processes progressively decrease in length to the fifteenth or sixteenth caudal vertebra, and then disappear. The neural spines progressively decrease in every dimension, save length, which is rather increased as far as the twenty-second or twenty-third vertebra, beyond which they begin again to shorten, and finally subside in the terminal vertebræ of the tail.

The caudal hæmapophyses coalesce at their lower or distal ends, from which a spinous process is prolonged downwards and backwards; this grows shorter towards the end of the tail, but is compressed and somewhat expanded antero-posteriorly. The hæmal arch so constituted has received the name of 'chevron bone.'

A side view of the body of a middle caudal vertebra of the *Crocodilus toliapicus* is given in T. III, fig. 8, and an under view of the same in fig. 9, showing the two hypapophysial ridges extending from the articular facets for the hæmapophyses at one end to the other end of the centrum.

The segments of the endo-skeleton composing the skull are more modified than those of the pelvis; but just as the vertebral pattern is best preserved in the neural arches of the pelvis, which are called collectively 'sacrum,' so, also, is it in the same arches of the skull, which are called collectively 'cranium.' The elements of which these cranial arches are composed preserve, moreover, their primitive or normal individuality more completely than in any of the vertebræ of the trunk, except the atlas, and consequently the archetypal character can be more completely demonstrated.

In fossil *Crocodilia*, and many other reptiles, the bones of the head are very liable from this cause to a greater extent of dislocation and separation than happens to the skull of the warm-blooded animal, in which a greater proportion of those primitive bones coalesce with age. It not unfrequently happens that detached bones of the skull of a reptile are found fossil, and the usually much modified form of these vertebral elements renders their determination difficult. In order to diminish this difficulty, especially as the bones of the cranium are least familiar to the palæontologist in their detached state, I have subjoined a side view of them, fig. 9, nearly as they are arranged in the formation of the successive natural segments of the skull. Such figures are the more necessary in the present state of anatomy and palæontology, since the illustrations of the osteology of the crocodile which have hitherto been prefixed to the descriptions of the fossil remains of the Reptilian class, as, e. g., in the great work of Cuvier, include only figures of the bones in question as they are naturally combined together in the entire skull.

For the anatomical description and determination of the individual bones, as constituent elements of the vertebral segments of the head, I must refer the reader to my work 'On the Archetype of the Vertebrate Skeleton,' pp. 115-25, figs. 18-21, and pl. 2, fig. 3; and here limit myself to an exemplification of the natural arrangement and names of the bones according to the letters and numbers in figure 9.

The bones of the head of the Crocodiles, as of all other vertebrate animals, are primarily classified into those of

The Endo-skeleton,
The Splanchno-skeleton, and
The Exo-skeleton.

Fig. .9

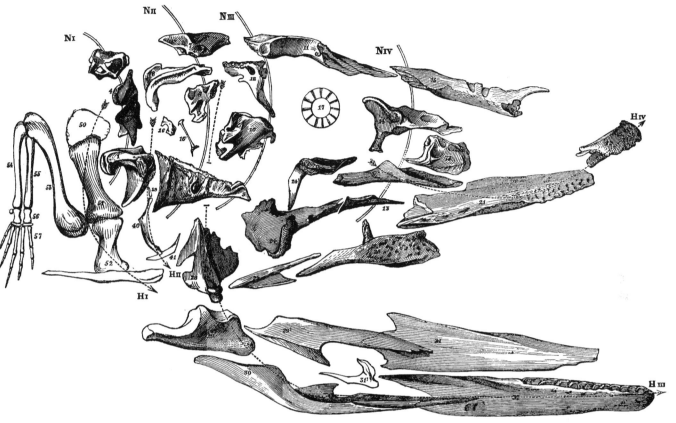

Disarticulated bones of the Skull of an Alligator, N ı to ıv the neural arches; H ı to ıv the hæmal arches and appendages.

The bones of the *endo-skeleton* of the head form naturally four segments, called

Occipital vertebra, N ı, H ı ;
Parietal vertebra, N ıı, H ıı ;
Frontal vertebra, N ııı, H ııı ; } Fig. 9.
Nasal vertebra, N ıv, H ıv.

These segments are subdivided into the neural arches, called

Epencephalic arch (1 basioccipital, 2 exoccipital, 3 superoccipital, 4 connate paroccipital) ;

Mesencephalic arch (5 basisphenoid, 6 alisphenoid, 7 parietal, 8 mastoid) ;

Prosencephalic arch (9 presphenoid, 10 orbitosphenoid, 11 frontal, 12 postfrontal) ;

Rhinencephalic arch (13 vomer, 14 prefrontal, 15 nasal) :

and into the hæmal arches and their appendages, called

Maxillary arch (20 palatine, 21 maxillary, 22 premaxillary) and appendages (24 pterygoid, 24' ectopterygoid, 26 malar, 27 squamosal) ;

Mandibular arch (28 tympanic, 29—32 mandible) ;

3

Hyoidean arch (39 epihyal, 40 ceratohyal, 41 basihyal) ;

Scapular arch (50 suprascapula, 51 scapula, 52 coracoid) and appendages
(53—58 bones of fore-limb).

The bones of the *splanchno-skeleton*, are

The petrosal (16) and otosteals (16′) ;

The sclerotals (17) which in most retain their primitive histological condition
as fibrous membrane.

The turbinals (18 and 19) and teeth.

The bones of the *exo-skeleton*, are

The lacrymals (73).

The superorbitals (present in *Alligator sclerops*).

To the foregoing brief analysis of the constituent parts of the framework of the *Crocodilia*, which are petrifiable or conservable in a fossil state, and from the study and comparison of which we have to gain our insight into the nature and affinities of the extinct Reptiles, it seems here only requisite to add a few observations on the characteristic mode in which the bones are associated together in certain parts of the skeleton in the present order, and especially in the skull.

With regard to the trunk, the *Crocodilia* are distinguished from the *Lacertilia* and from all other existing orders of Reptiles, by the articulation of the vertebral ribs (pleurapophyses) in the cervical and anterior part of the dorsal segments by a head and tubercle to a parapophysis and diapophysis. As this double joint is associated with a double ventricle of the heart, and as the single articulation of every rib in other Reptiles is associated with a single ventricle of the heart, we may infer a like difference in the structure of the central organ of circulation in the extinct reptiles, manifesting the above-defined modifications in the proximal joints of the ribs.

The sacrum consists of two vertebræ only, in *Crocodilia* as in *Lacertilia*: they are modified in the present order, as before described, p. 14.

The skull consists, as above defined, of four segments. The hinder or occipital surface of the skull presents, in the *Crocodilia* as in the *Lacertilia*, a single convex occipital condyle, formed principally by the basioccipital, and not showing the trefoil character which it bears in the *Chelonia* (Part I, T. XV, fig. 4), in which the exoccipitals contribute equal shares to its formation. In the *Batrachia*, the exoccipitals exclusively form the joint with the atlas, and there are accordingly two condyles. The occipital region of the crocodilian skull is remarkable for its solidity and complete ossification, and for the great extent of the surface which descends below the condyle. (T. VI, fig. 2.) In the *Lacertilia*, a wide vacuity is left between the mastoid, exoccipital, and paroccipital: but in the *Crocodilia* this is reduced to the small depressions or foramina near 3, fig. 2, T. VI. The tympanic pedicles (28) extend outwards and downwards, firmly wedged between the paroccipital, mastoid, and squamosal; in the Lacertians these pedicles are suspended vertically from the point of union of the mastoid and paroccipital.

The chief foramen in the occipital region is that called 'foramen magnum' (between 2 and 2, in fig. 2, T. VI), through which the nervous axis is continued from the skull. On each side of the foramen magnum is a small hole, called 'precondyloid foramen,' for the exit of the hypoglossal nerve. External to this is a larger foramen, marked *n* in fig. 2, for the transmission of the nervus vagus and a vein. Below this is the 'carotid foramen' *c*. All these are perforated in the exoccipital. Below the condyle there is usually a foramen, and sometimes two, for the transmission of blood-vessels. Lower down, at the suture between the basioccipital and basisphenoid, is a larger and more constant median foramen, indicated by the dotted line from *e t;* it is the bony outlet of a median system of eustachian tubes, peculiar to the *Crocodilia*. On each side of the median eustachian foramen, and in the same suture, is a smaller foramen, which is the bony orifice of the ordinary lateral eustachian tube. The membranous continuations of the lateral eustachian tubes unite with the shorter continuation from the median tube, and all three terminate by a common valvular aperture, upon the middle line of the faucial palate, behind the posterior or palatal nostril. The large, bony aperture of this nostril is formed by the pterygoids (24 in fig. 2). The carotid canal, *c*, opens by a short bony tube into the tympanic cavity, and is described as the 'eustachian canal' in the 'Leçons d'Anatomie comparée' of Cuvier. The artery crosses the tympanic cavity, and enters a bony canal at its fore part, which conducts to the 'sella turcica' in the interior of the cranium.

The median eustachian foramen is described by Cuvier as the 'arterial foramen,'[*] the canal from which divides and terminates in the 'sella turcica.'[†] By MM. Bronn, Kaup, and De Blainville, the median eustachian foramen is contended to be the bony aperture of the posterior nostrils.[‡]

The results of the dissections and injections of recent Crocodiles and Alligators, by which I have been able to rectify the discrepant opinions regarding the carotid, eustachian, and naso-palatal foramina, and which have led to the discovery of a third median eustachian canal, or rather system of canals, between the tympanic cavities and fauces, peculiar to the Crocodilian Reptiles, are given in detail in the 'Philosophical Transactions' for 1850. The complexity of the superadded system has doubtless chiefly contributed to mislead the justly-esteemed authorities who have believed that they saw in it characters of the carotid canals or of the posterior nasal passages. The eustachian apparatus in the *Crocodilia* may be briefly described as follows: From the floor of each tympanic cavity two air-passages are continued; the canal from the fore part of the cavity extends downwards, backwards, and inwards, in the basisphenoid, which

[*] Ossemens Fossiles, tom. v, pt. ii, p. 133.

[†] Ib. p. 78.

[‡] Abhandlungen über die Gavialärtigen Reptilien der Lias-formation, folio, 1841, pp. 12, 16, 44.

unites with its fellow from the opposite tympanum, to form a short median canal, which descends backwards to the suture between the basisphenoid and the basioccipital, where it joins the median canal formed by the union of the two air-passages from the back part of the floor of the tympanum, which traverse the basioccipital. The common canal formed by the junction of the two median canals descends along the suture to the median foramen *e t*, fig. 2, T. VI. The air-passage from the back part of the tympanum, which traverses the basioccipital, swells out into a rhomboidal sinus in its convergent course towards its fellow, and from this sinus is continued the normal lateral eustachian canal, which, on each side, terminates below in the small aperture, external to the median eustachian foramen.

That part of the outer surface of the skull which is covered by the common integument is more or less sculptured with wrinkles and pits in the *Crocodilia*: the modifications of this pattern are shown in T. I, fig. 1, in the nilotic Crocodile, and in T. VI, in the eocene Crocodile from Hordwell. The flat platform of the upper surface of the cranium is perforated by two large apertures, surrounded by the bones numbered 7, 8, 11, 12; these apertures are the upper outlets of the temporal fossæ, divided from the lower and lateral outlets by the conjoined prolongations of the mastoid 8 and postfrontal 12: if ossification were continued thence to the parietal 7, the temporal fossæ would be roofed over by bone, as in the *Chelones*. In old Crocodiles and Alligators there is an approximation to this structure, and the upper temporal apertures are much diminished in size. In the Gavials (T. XI, fig. 1 *a*) they remain more widely open, and, in the fossil Gavials of the secondary strata, they are still wider, as seen in T. XI, fig. 2 *a*; by which the structure of the cranium approaches more nearly to that of the Lacertian reptiles, where the temporal fossa is either not divided into an upper and lateral outlet, or is bridged over by a very slender longitudinal bar from the postfrontal to the mastoid. The lateral outlets of the temporal fossæ (T. VI, fig. 1) are divided from the orbits by a bar of bone developed from the postfrontal (12) and malar (26), and against the inner side of the base of which the ectopterygoid abuts; the posterior boundary of the fossa is made by the tympanic (28) and squamosal (27). The orbits, having the postfronto-malar bar (12, 26) behind, are surrounded in the rest of their circumference by the frontal (11), the prefrontal (14), the lachrymal (73), and the malar (26). The supraorbital or palpebral ossicle is rarely preserved in fossil specimens.

The facial or rostral part of the skull anterior to the orbit, is of great extent, broad and flat in the Alligators and some Crocodiles, narrower, rounder, and longer in other Crocodiles, always most narrow, cylindrical, and elongated in the Gavials. The anterior or external nostril is single, and is perforated in the middle of the anterior terminal expansion of the upper jaw. This expansion is least marked in the broad-headed species (compare T. VI, fig. 1, with T. III, fig. 1); in existing Crocodiles and Alligators the points of the nasal bones penetrate its hind border, as at 15, fig. 1,

T. I.   In the Gavials (T. XI, fig. 1 *a*) the nasals (*n*) terminate a long way from the nostril.   The *Crocodilia* resemble the *Chelonia* in the single median nostril.*   In the *Lacertilia* there is a pair of nostrils, one on each side the median plane, which is occupied by a bridge of bone extending from the usually single premaxillary to the nasals.   The plane of the single nostril is almost horizontal in all existing and tertiary *Crocodilia*.

On the inferior or palatal surface of the skull (T. VII, fig. 2), the most anterior aperture is the circular prepalatal foramen surrounded by the premaxillaries 22; then follows an extensive smooth, horizontal, bony plate, formed by the premaxillaries (22), the maxillaries (21), and the palatines (20).   The postpalatal apertures are always large in the *Crocodilia*, and are bounded by the palatines (20), maxillaries (21), pterygoids (24), and ectopterygoids (25).   The posterior aperture of the nostril is formed wholly by the pterygoids; it is shown in T. VI, fig. 3, between the bones marked 24.   Behind it is the median and lateral eustachian foramen already described, as belonging rather to the posterior region of the head.

The scapulo-coracoid arch, both elements (fig. 9, 51, 52) of which retain the form of strong and thick vertebral and sternal ribs in the crocodile, is applied in the skeleton of that animal over the anterior thoracic hæmal arches (T. XI).   Viewed as a more robust hæmal arch, it is obviously out of place in reference to the rest of its vertebral segment. If we seek to determine that segment by the mode in which we restore to their centrums the less displaced neural arches of the antecedent vertebræ of the cranium or in the sacrum of the bird,† we proceed to examine the vertebræ before and behind the displaced arch, with the view to discover the one which needs it, in order to be made typically complete.   Finding no centrum and neural arch without its pleurapophyses from the scapula to the pelvis, we give up our search in that direction; and in the opposite direction we find no vertebra without its ribs until we reach the occiput: there we have centrum and neural arch, with coalesced parapophyses—the elements answering to those included in the arch N I, fig. 9—but without the arch H I; which arch can only be supplied, without destroying the typical completeness of antecedent cranial segments, by a restoration of the bones 50—52, to the place which they naturally occupy in the skeleton of the fish.   And since anatomists are generally agreed to regard the bones 50—52 in the crocodile as specially homologous with those so numbered in the fish,‡ we must conclude that they are likewise homologous in a higher sense; that in the fish, the scapulo-coracoid arch is in its natural or typical position, whereas in the crocodile it has been displaced for a special purpose. Thus, agreeably with a general principle, we perceive that, as the lower vertebrate

---

* In a skeleton of the *Alligator lucius* in the Museum of the Royal College of Surgeons, a slender bar of bone is continued from the nasals to the premaxillary, across the median nasal aperture, as it is in the skull of the same species figured in the 'Ossemens Fossiles,' tom. v, pt. ii, pl. i, fig. 8.

† See 'On the Archetype and Homologies of the Vertebrate Skeleton,' p. 117, p. 159.

‡ Op. cit., fig. 5, p. 17.

animal illustrates the closer adhesion to the archetype by the natural articulation of the scapulo-coracoid arch to the occiput, so the higher vertebrate manifests the superior influence of the antagonising power of adaptive modification by the removal of that arch from its proper segment.

The anthropotomist, by his mode of counting and defining the dorsal vertebræ and ribs, admits, unconsciously perhaps, the important principle in general homology which is here exemplified, and which, pursued to its legitimate consequences and further applied, demonstrates that the scapula is the modified rib of that centrum and neural arch which he calls the 'occipital bone,' and that the change of place which chiefly masks that relation (for a very elementary acquaintance with comparative anatomy shows how little mere form and proportion affect the homological characters of bones) differs only in extent and not in kind from the modification which makes a minor amount of comparative observation requisite, in order to determine the relation of the shifted dorsal rib to its proper centrum in the human skeleton.

With reference, therefore, to the occipital vertebra of the crocodile, if the comparatively well-developed and permanently distinct ribs of all the cervical vertebræ prove the scapular arch to belong to none of those segments, and, if that hæmal arch be required to complete the occipital segment, which it actually does complete in fishes, then the same conclusion must apply to the same arch in other animals, and we must regard the occipital vertebra of the tortoise as completed below by its scapulo-coracoid arch and not, as Bojanus supposed, by its hyoidean arch.*

Having thus endeavoured to show what the scapular arch of the crocodile is, I proceed to point out the characteristic form of its chief elements. The upper and principal part of the scapula (51, fig. 9) is flattened, and gradually becomes narrower to the part called its neck, which is rounded, bent inwards, and then suddenly expanded to form a rough articular surface for the coracoid, and a portion of a smoother surface for the shoulder-joint. The contiguous end of the coracoid (52) presents a similar form, having not only the rough surface for its junction with the scapula, but contributing, also, one half of the cavity for the head of the humerus. It is perforated near the interspace between these two surfaces. As it recedes from them, it contracts, then expands and becomes flattened, terminating in a somewhat broader margin than the base of the

---

* Anatome Testudinis Europæa, fol., 1819, p. 44. Geoffroy St. Hilaire selected the opercular and subopercular bones to form the inverted arch of his seventh (occipital) cranial vertebra, and took no account of the instructive natural connexions and relative position of the hyoidean and scapular arches in fishes. With regard to the scapular arch, he alludes to its articulation with the skull in the lowest of the vertebrate classes as an 'amalgame inattendue' (*Anatomie Philosophique*, p. 481) : and elsewhere describes it as a "disposition véritablement très singulière, et que le manque absolu du cou et une combinaison des pièces du sternum avec celles de la tête pouvoient seules rendre possible."—*Annales du Muséum*, ix, p. 361. A due appreciation of the law of vegetative uniformity or repetition, and of the ratio of its prevalence and power to the grade of organization of the species, was, perhaps, essential in order to discern the true signification of the connexion of the scapular arch in fishes.

scapula, which margin is morticed into a groove at the anterior border of the broad rhomboidal cartilage continued beyond the ossified part of the manubrium, which forms the key-bone of the scapular arch. The anterior locomotive extremity is the diverging appendage of the arch, under one of its numerous modes and grades of development.*

The proximal element of this appendage or that nearest the arch, is called the 'humerus' (53, fig. 9): its head is subcompressed and convex; its shaft bent in two directions, with a deltoid crest developed from its upper and fore part; its distal end is transversely extended, and divided anteriorly into two condyles. The shaft of this bone has a medullary cavity, but relatively smaller than in the mammalian humerus.

The second segment of the limb consists of two bones: the larger one (54) is called the 'ulna:' it articulates with the outer condyle of the humerus by an oval facet, the thick convex border of which swells a little out behind, and forms a kind of rudimental 'olecranon;' the shaft of the ulna is compressed transversely, and curves slightly outwards; the distal end is much less than the proximal one, and is most produced at the radial side.

The radius (55) has an oval head; its shaft is cylindrical; its distal end oblong and subcompressed.

The small bones (56) which intervene between these and the row of five longer bones, are called 'carpals:' they are four in number in the Crocodilia. One seems to be a continuation of the radius, another of the ulna; these two are the principal carpals; they are compressed in the middle and expanded at their two extremities; that on the radial side of the wrist is the largest. A third small ossicle projects slightly backwards from the proximal end of the ulnar metacarpal: it answers to the bone called 'pisiforme' in the human wrist. The fourth ossicle is interposed between the ulnar carpal and the metacarpals of the three ulnar digits.

These five terminal jointed rays of the appendage are counted from the radial to the ulnar side, and have received special names: the first is called 'pollex,' the second 'index,' the third 'medius,' the fourth 'annularis,' and the fifth 'minimus.' The first joint of each digit is called 'metacarpal;' the others are termed 'phalanx.' In the Crocodilia the pollex has two phalanges, the index three, the medius four, the annularis four, and the minimus three. The terminal phalanges, which are modified to support claws, are called 'ungual' phalanges.

As the above-described bones of the scapular extremity are developments of the appendage of the scapular arch, which is the hæmal arch of the occipital vertebra, it follows, that, like the branchiostegal rays and opercular bones in fishes, they are essentially bones of the head.

The diverging appendage of the pelvic arch being a repetition of the same element

* See my Discourse 'On the Nature of Limbs,' 8vo, Van Voorst, 1849, pp. 64-70.

as the appendage of the scapular arch modified and developed for a similar office, a close resemblance is maintained in the subdivisions of the framework of both limbs. The first bone of the pelvic limb, called the 'femur,' is longer than the humerus, and, like it, presents an enlargement of both extremities, with a double curvature of the intervening shaft, but the directions are the reverse of those of the humerus, as may be seen in T. XI, where the upper or proximal half of the femur is concave, and the distal half convex, anteriorly. The head of the femur is compressed from side to side, not from before backwards as in the humerus; a pyramidal protuberance from the inner surface of its upper fourth represents a 'trochanter;' the distal end is expanded transversely, and divided at its back part into two condyles.

The next segment or 'leg,' includes, like the corresponding segment of the forelimb called 'forearm,' two bones. The largest of these is the 'tibia,' and answers to the radius. It presents a large, triangular head to the femur; it terminates below by an oblique crescent with a convex surface.

The 'fibula' is much compressed above; its shaft is slender and cylindrical, its lower end is enlarged and triangular.

All these long bones have a narrow medullary cavity.

The group of small bones which succeed those of the leg, are the tarsals; they are four in number, and have each a special name. The 'astragalus' articulates with the tibia, and supports the first and part of the second toe. It is figured in Cuvier's 'Ossemens Fossiles,' tom. v, pt. ii, pl. iv, figs. 19 *A, B, C, D.* The 'calcaneum' intervenes between the fibula and the ossicle supporting the two outer toes; it has a short but strong posterior tuberosity.

The ossicle referred to represents the bone called 'cuboid' in the human tarsus. A smaller ossicle, wedged between the astragalus and the metatarsals of the second and third toes is the 'ectocuneiform.'

Four toes only are normally developed in the hind-foot of the *Crocodilia ;* the fifth is represented by a stunted rudiment of its metatarsal, which is articulated to the cuboid and to the base of the fourth metatarsal.

The four normal metatarsals are much longer than the corresponding metacarpals. That of the first or innermost toe is the shortest and strongest; it supports two phalanges. The other three metatarsals are of nearly equal length, but progressively diminish in thickness from the second to the fourth. The second metatarsal supports three phalanges; the third four; and the fourth also has four phalanges, but does not support a claw. The fifth digit is represented by a rudiment of its metatarsal in the form of a flattened triangular plate of bone, attached to the outer side of the cuboid, and slightly curved at its pointed and prominent end.

In the skull of the Crocodile, as of most other vertebrates, there are intercalated a few bones, or ossified parts of special organs, which, as is shown in the classed Table of the bones of the head, do not belong to the vertebral system of bones.

The bone anterior to the orbit, marked 73 in fig. 9, and in T. I and T. VI is perforated at its orbital border by the duct of the lachrymal gland, whence it is termed the 'lachryma' bone,' and its facial part extends forwards between the bones marked 14, 15, 21, and 26 in the same plates. In many Crocodilia there is a bone at the upper border of the orbit, which extends into the substance of the upper eyelid; it is called 'superorbital.' In the *Crocodilus palpebrosus* there are two of these ossicles.

Both the lachrymal and superorbital bones answer to a series of bones found commonly in fishes, and called 'suborbitals' and 'superorbitals.' The lachrymal is the most anterior of the suborbital series, and is the largest in fishes; it is also the most constant in the vertebrate series, and is grooved or perforated by a mucous duct. These ossicles appertain to the dermal or muco-dermal system or 'exoskeleton.'

The little slender bone, marked 16′ in fig. 9, has one of its extremities in the form of a long, narrow, elliptic plate, which is applied to the 'fenestra ovalis' of the internal ear; from this plate extends a long and slender bony stem, which grows somewhat cartilaginous, expands and bends down, as it approaches the tympanum or ear-drum, to which it is attached.

The cartilaginous capsule of the labyrinth or internal ear is partially ossified by sinuous plates of bone connate with the neurapophyses (fig. 10, 2 and 6), between which that organ is lodged; I apply the term 'petrosal' to the principal and most independent of those ossifications of the ear-capsule, to that, e. g., which retains

some mobility after it has contracted a partial anchylosis to the exoccipital (2), and which appears upon the inner surface of the cranial walls at the part marked 16 in the subjoined Cut 10, between 2 and 6. It is the only independent bone on that surface of the cranium which, in my opinion, answers to the 'petrous portion of the temporal' in human anatomy, and to which the term 'rocher'

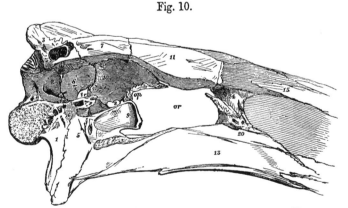

Fig. 10.

Vertical longitudinal section of the cranium of a Crocodile (*Crocodilus acutus*).

can be properly applied, in the language of the French comparative anatomists. Cuvier, however, restricts that name to the 'alisphenoid' (6, figs. 9, 10) in the Crocodiles.

The ossicles, (16 and 16′, fig. 9), together with the partial ossifications in the sclerotic capsule of the organ of sight, (17, fig. 9)—always more distinct in *Chelonia* than in *Crocodilia*— belong to that category of visceral bones to which the term 'splanchnoskeleton' has been given; they also are foreign to the true vertebrate system of the skeleton.

The teeth.—The most readily recognisable character by which the existing Crocodilians are classified and grouped in appropriate genera, are derived from modifications of the dental system.

In the Caimans (genus *Alligator*) the teeth vary in number from $\frac{18-18}{18-18}$ to $\frac{22-22}{22-22}$: the fourth tooth of the lower jaw is *received into a cavity* of the alveolar surface of the upper jaw, where it is concealed when the mouth is shut. In T. VIII, fig. 2, these pits are shown behind the last premaxillary tooth *e*, in an Eocene Alligator from Hordwell. In old individuals of the existing species of Alligator, the upper jaw is perforated by the large inferior teeth in question, and the fossæ are converted into foramina.

In the Crocodiles (genus *Crocodilus*) the fourth tooth in the lower jaw is *received into a notch* excavated in the side of the alveolar border of the upper jaw, as in fig. 1, T. VIII, behind the tooth *e*, and is visible externally when the mouth is closed, as in T. VII, fig. 1. In most Crocodiles, also, the first tooth in the lower jaw perforates the premaxillary bone when the mouth is closed, as in T. I, between the teeth marked *a* and *b*.

In the two preceding genera the alveolar borders of the jaw have an uneven or wavy contour, and the teeth are of an unequal size.

In the Gavials, (genus *Gavialis*) the teeth are nearly equal in size and similar in form in both jaws, and the first as well as the fourth tooth in the lower jaw, passes into a groove in the margin of the upper jaw when the mouth is closed, T. XI.

In the Alligators and Crocodiles the teeth are more unequal in size, and less regular in arrangement, and more diversified in form than in the Gavials: witness the strong thick conical laniary teeth at the fore part of the jaw, as shown in T. VII and T. III, fig. 6, as contrasted with the blunt mammillate summits of the posterior teeth, as shown in T. V, fig. 12. The teeth of the Gavial are subequal, most of them are long, slender, pointed, subcompressed from before backwards, with a trenchant edge on the right and left sides, between which a few faint longitudinal ridges traverse the basal part of the enamelled crown.

The teeth of both the existing and extinct Crocodilian reptiles consist of a body of compact dentine forming a crown covered by a coat of enamel, and a root invested by a moderately thick layer of cement. The root slightly enlarges, or maintains the same breadth to its base, which is deeply excavated by a conical pulp-cavity extending into the crown, and is commonly either perforated or notched at its concave or inner side.

The dentinal tubes in the crown of a fully-developed tooth form short curvatures at their commencement at the surface of the pulp-cavity, and then proceed nearly straight to the periphery of the crown; they very soon bifurcate, the divisions slightly diverging; then continuing their course with gentle parallel undulations, they subdivide near the enamel, and terminate in fine and irregular branches, which anastomose generally by the medium of cells. The dentinal tubes send off from both sides, throughout their progress, minute branches into the intervening substance, and terminate in the dentinal cells. These cells are subhexagonal, about $\frac{1}{800}$ of an inch in diameter, and are traversed by from ten to fourteen of the dentinal tubes; they are usually arranged in planes parallel with the periphery of the crown, near which they are most conspicuous, and towards which their best defined outline is directed: they combine with the parallel curvatures of the dentinal tubes to form the striæ, visible in sections of the teeth by the naked eye, which cause the stratified appearance of the dentine as if it were composed of a succession of superimposed cones. The diameter of the dentinal tube before the first bifurcation is $\frac{1}{12000}$th of an inch, both the trunks and bifurcations of the tubes have interspaces equal to four of their respective diameters.

The enamel viewed in a transverse section of the crown presents some delicate striæ parallel with its surface, whilst the appearance of fibres vertical to that surface is only to be detected, and these faintly, on the fractured edge. It is a very compact and dense substance; the dark brownish tint is strongly marked in the middle of the enamel when viewed by transmitted light.

The cells with which the fine tubes of the basal cement communicate, are oblong, about $\frac{1}{2000}$th of an inch across their long axis, which is transverse to that of the tooth; the inter-communicating tubes, which radiate from the cells, giving them a stellate figure. I have entered into these particulars of the microscopic texture of the teeth of the Crocodile because it will be seen in the sequel that important modifications of the dental tissues characterise some of the extinct *Reptilia*.

In the black Alligator of Guiana the first fourteen teeth of the lower jaw are implanted in distinct sockets, the remaining posterior teeth are lodged close together in a continuous groove, in which the divisions for sockets are faintly indicated by vertical ridges, as in the jaws of the Ichthyosaurs. A thin compact floor of bone separates this groove, and the sockets anterior to it, from the large cavity of the ramus of the jaw; it is pierced by blood-vessels for the supply of the pulps of the growing teeth and the vascular dentiparous membrane which lines the alveolar cavities.

The tooth-germ is developed from the membrane covering the angle between the floor and the inner wall of the socket. It becomes in this situation completely enveloped by its capsule, and an enamel-organ is formed at the inner surface of the capsule before the young tooth penetrates the interior of the pulp-cavity of its predecessor.

The matrix of the young growing tooth affects, by its pressure, the inner wall of the socket, and forms for itself a shallow recess; at the same time it attacks the side of the base of the contained tooth; then, gaining a more extensive attachment by its basis and increased size, it penetrates the large pulp-cavity of the previously formed tooth, either by a circular or semicircular perforation. The size of the calcified part of the tooth matrix which has produced the corresponding absorption of the previously formed tooth on the one side, and of the alveolar process on the other, is represented in the second exposed alveolus of the portion of jaw figured in Pl. 75, fig. 4, of my 'Odontography,' the tooth marked *a* in that figure, having been displaced and turned round to show the effects of the stimulus of the pressure. The size of the perforation in the tooth, and of the depression in the jaw, proves them to have been, in great part, caused by the soft matrix, exciting dissolution and absorbent action, and not by mere mechanical force. The resistance of the wall of the pulp-cavity having been thus overcome, the growing tooth and its matrix recede from the temporary alveolar depression, and sink into the substance of the pulp contained in the cavity of the fully-formed tooth. As the new tooth grows, the pulp of the old one is removed; the old tooth itself is next attacked, and the crown being undermined by the absorption of the inner surface of its base, may be broken off by a slight external force, when the point of the new tooth is exposed.

The new tooth disembarrasses itself of the cylindrical base of its predecessor, with which it is sheathed, by maintaining the excitement of the absorbent process so long as the cement of the old fang retains any vital connexion with the periosteum of the socket; but the frail remains of the old cylinder, thus reduced, are sometimes lifted off the socket upon the crown of the new tooth, when they are speedily removed by the action of the jaws. This is, however, the only part of the process which is immediately produced by mechanical force: an attentive observation of the more important previous stages of growth, teaches that the pressure of the growing tooth operates upon the one to be displaced only through the medium of the vital dissolvent and absorbent action which it has excited.

Most of the stages in the development and succession of the teeth of the Crocodiles are described by Cuvier* with his wonted clearness and accuracy; but the mechanical explanation of the expulsion of the old tooth, which Cuvier adopts from M. Tenon, is opposed by the disproportion of the hard part of the new tooth to the vacuity in the walls of the old one, and by the fact that the matter impressing—viz. the uncalcified part of the tooth-matrix—is less dense than the part impressed.

No sooner has the young tooth penetrated the interior of the old one, than another germ begins to be developed from the angle between the base of the young tooth and the inner alveolar process, or in the same relative position as that in which its immediate predecessor began to rise, and the processes of succession and displacement are carried on, uninterruptedly, throughout the long life of these cold-blooded carnivorous reptiles.

From the period of exclusion from the egg, the teeth of the crocodile succeed each other in the vertical direction; none are added from behind forwards, like the true

* Op. cit., pp. 90-3.

molars in Mammalia. It follows, therefore, that the number of the teeth of the cro-codile is as great when it first sees the light as when it has acquired its full size; and, owing to the rapidity of the succession, the cavity at the base of the fully-formed tooth is never consolidated.

The fossil jaws of the extinct Crocodilians demonstrate that the same law regulated the succession of the teeth at the ancient epochs when those highly organized reptiles prevailed in greatest numbers, and under the most varied generic and specific modi-fications, as at the present period, when they are reduced to a single family, composed of so few and slightly varied species, as to have constituted in the *Systema Naturæ* of Linnæus, a small fraction of the genus *Lacerta*.

CROCODILUS TOLIAPICUS, *Owen.* Tab. II, fig. 1, and Tab. II *A*.

Syn. CROCODILE DE SHEPPY (?), *Cuvier.* Ossemens Fossiles, 4to, tom. v, pt. ii, p. 165.
CROCODILUS SPENCERI, *Buckland.* Bridgewater Treatise, vol. i, p. 251. "Crocodile with a short and broad snout." Vol. ii, p. 36, pl. 25', fig. 1.
— — *Owen.* Reports of the British Association, 1841, p. 65.

In proceeding to the comparison, and preparing for the description of the British fossil *Crocodilia*, I endeavoured, in the first place, to obtain the bones of the species which now exists in a locality nearest to Great Britain, and also of an individual of that same species which had lived at a remote period; and I have been favoured by the kindness of my esteemed friend Philip Duncan, Esq., Fellow of New College, Oxford, and Conservator of the Ashmolean Museum, with the opportunity of examining the bones of a mummified Crocodile from a sarcophagus at Thebes, in that collection at Oxford. Two views of the skull of this old Egyptian Crocodile are given in T. I. The total length of the skull from the bone marked 28 to the end of 22, is twice the breadth of the back part of the skull. The upper apertures of the temporal fossæ are subcircular; the point of the squamosal (27) projects into the lateral aperture. The breadth of the back part of the sculptured cranial platform (8,8), is less by one fourth than the breadth of the skull anterior to the orbits. The breadth of the interorbital space is nearly equal to the transverse diameter of the orbit. The points of the nasals (15) project into the external nostril. The postpalatal apertures reach as far forwards as the seventh tooth, counting from the hindmost; there are nineteen alveoli on each side of the upper jaw, the five anterior teeth being lodged in the premaxillary, which is perforated by the first tooth of the lower jaw.

Geoffroy St. Hilaire has applied the old Egyptian name Σοῦχος to the mummified Crocodiles of that country; but there is no good specific character which distinguishes them from the modern Crocodiles of the Nile, to which Cuvier has given the name of *Crocodilus vulgaris.*

Cuvier appears to have first called the attention of palæontologists to the remains of *Crocodilia* in the Eocene clay forming the Isle of Sheppy, in the last volume of the second edition of his great work on the 'Ossemens Fossiles,' p. 165, 1824. He there specifies a third cervical vertebra, which was obtained by M. G. A. Deluc, at Sheppy, and of which Cuvier made a drawing at Geneva; he says it much resembles the corresponding vertebra in one of our living Crocodiles, and might have come from an individual about five feet in length. "M. Deluc," he adds, "found very near it a much smaller vertebra, which I recognised as belonging to a monitor or some allied genus."*

Our knowledge of the Eocene Crocodiles of Sheppy received a remarkable accession at the publication of the highly interesting and instructive 'Bridgewater Treatise' of Dr. Buckland, in which he states that " true Crocodiles, with a short and broad snout, like that of the Caiman and the Alligator, appear, for the first time, in strata of the tertiary periods, in which the remains of mammalia abound. . . . One of these," he adds, " found by Mr. Spencer in the London Clay of the Isle of Sheppy, is engraved Pl. 25', fig. 1," and the name ' *Crocodilus Spenceri'* is appended to that figure.

In preparing my 'Report on British Fossil Reptiles' for the British Association in 1841, I examined the original specimen figured by Dr. Buckland, in which unfortunately the end of the snout with the intermaxillaries and an indeterminate proportion of the maxillaries having been broken off and lost, no exact idea could be formed of the proportions of the facial or rostral part of the skull.

In a larger specimen of the fossil skull of a Crocodile from Sheppy, in the British Museum, T. II *A*, the whole of the upper, as well as the lower jaw, were preserved, and as the proportions of the snout agreed with those of some true Crocodiles, and differed in an equal degree with those species from the Gavial; and as, like the Crocodiles and Caimans, it presented the more important distinction of a different composition of that part of the skull, I retained for the specimen in that 'Report' the name of *Crocodilus Spenceri*, proposed by the author of the Bridgewater Treatise for the Sheppy Crocodile, so differing from the Gavial.

The able keeper of the Mineralogical Department of the British Museum, Charles König, K.H., F.R.S., to whom I am indebted for every facility in describing and figuring this specimen, has suggested that the name by which Baron Cuvier first indicated the existence of a true Crocodile in the Eocene clay of Sheppy, should have the priority, and I adopt, therefore, the name *Crocodilus toliapicus,* which he has attached to the specimen in question, and with the more readiness since I have now reason to doubt whether the mutilated cranium, figured in the 'Bridgewater Treatise,' belongs to the same species.

---

* Could this have been a vertebra of the large serpent, which I have subsequently described under the name of *Palæophis?* I have not as yet met with a single lacertian vertebra from Sheppy. If the collection of M. Deluc be still preserved at Geneva, the vertebra in question might be compared with the figures of the *Palæophus toliapicus,* 'Ophidia,' T. XV.

The more entire fossil skull in question presents the following dimensions:

	Feet.	Inches.	Lines.
Total length from the hindmost part of the lower jaw .	2	2	0
Breadth between the articular ends of the tympanics .	0	10	0
Do. across the orbits	0	7	6
Do. of the intertemporal space	0	0	9
Do. of the interorbital space	0	1	4
From the articular end of the tympanic to the orbit .	0	8	6
From the occipital condyle to the orbit	0	7	0
From the orbit to the external nostril	0	14	0
Breadth of the cranium five inches in advance of the orbits	0	3	8
Do. across the external nostril	0	2	8
Depth of the lower jaw at the vacuity between the angular and surangular	0	3	6
Length of that vacuity	0	3	0
Breadth of the base of one of the larger maxillary teeth .	0	0	8

This remarkably fine fossil skull, which is figured one third of its natural size in T. II, fig. 1, presents proportions which come nearest to those of the *Crocodilus acutus*, being longer in proportion to its basal breadth than in the *Crocodilus Suchus*, in which the diameter between the articular ends of the tympanis (28) is just half the length of the entire skull. The interorbital space in the *Crocodilus toliapicus* is relatively narrower and flatter than in the *Croc. acutus* or *Croc. Suchus*, and the facial part of the skull becomes narrower before the expansion of the upper jaw, at the figure 15, than it does in either of those species. The narrow elongated nasals on which the figure 15 is placed, extend forwards to the external nostril (22), as in the true Crocodiles, and the alveolar border is festooned as is shown in the side view in T. II *A*. The teeth are $\frac{22-22}{20-20}$=84 in number: they are more uniform in size, and more regularly spaced than in the recent species above cited, and resemble in this respect the teeth of the *Crocodilus Schlegelii* of S. Müller, which is from Borneo. The extent of the symphysis of the lower jaw is greater in the *Crocodilus toliapicus* than in the *Croc. acutus*, and still greater than in the *Croc. Suchus;* the Sheppy species in this respect more nearly resembles the living species from Borneo above cited.

CROCODILUS CHAMPSOIDES, *Owen.*   Tab. III.   (Tab. II, fig. 2?)

Syn. CROCODILE DE SHEPPY (?), *Cuvier.* Loc. cit.
CROCOCILUS SPENCERI, or "Crocodile with a short and broad snout" (?) *Buckland.* Bridgewater Treatise, vol. ii, pl. xxv, fig. 1.

The fossil skull already described establishes the fact of the existence of a true Crocodile in the London Clay at Sheppy, but not of a species with a short and broad snout; the present specimen equally demonstrates the presence at the earliest period of the Tertiary geological epoch of *Crocodilia* with those modifications of the cranial and dental structure on which the characters of the restricted genus *Crocodilus* of modern Zoology are founded; but they are associated with a general form of the head which approaches more nearly to the Gavials than does that of the *Crocodilus toliapicus*, and which are most nearly paralleled amongst the known existing true Crocodiles by the *Crocodilus Schlegelii*. This Bornean species was, in fact, originally described as a new species of *Gavial*, but the nasal bones, as in the fossil from Sheppy figured in T. II, 15, extend to the hind border of the external nostril.

The fine subject of T. III, forms part of the collection of J. S. Bowerbank, Esq. F.R.S., which is well known for its rich and varied illustrations of the fossils of the Isle of Sheppy.

The following are some of its admeasurements:

	Feet.	Inches.	Lines.
Total length from the occipital condyle to the end of the premaxillaries	1	4	0
Breadth across the hinder angles of the supracranial platform	0	4	0
Do. across the orbits	0	5	0
Do. of the intertemporal space	0	0	4
Do. of the interorbital space	0	1	0
Do. across the external nostril	0	2	0
From the occipital condyle to the orbit . . .	0	3	4
From the orbit to the external nostril	0	10	0

The skull yielding the above dimensions is much smaller than that of the *Crocodilus toliapicus*, T. II; but it cannot have belonged to a younger individual of the same species, because, in existing Crocodiles, the part of the skull anterior to the orbits is proportionally shorter in the young than in the old individuals, as may be seen by comparing the figures which Cuvier has given of the skulls of three individuals of different ages of the *Crocodilus biporcatus*, in figures 4, 18, and 19, of plate 1 of the last volume of the 'Ossemens Fossiles,' 4to, 1824; whereas the part of the skull anterior to the orbits is relatively longer and more slender in the smaller fossil skull now described than in the larger one on which the species *Croc. toliapicus* is founded. We have, therefore, satisfactory proof that two species of true Crocodile existed during the deposition of the Eocene Clay at the actual mouth of the Thames, and have left their remains in that locality.

Their specific distinction is further illustrated by the different forms and proportions of particular parts of the skull. The alveolar border is more nearly straight; the transverse expansion of the maxillaries (21) is less, whilst that of the premaxillaries (22) is greater: the interorbital space is broader and more concave. The teeth are more uniform in size, are more regularly spaced, and are wider apart: they are, likewise, upon the whole, larger in proportion to the size of the jaw. Figure 5, T. III, shows the crown of a new tooth just emerging from the second socket of the maxillary bone of the natural size; figure 6 is the fourth tooth of the premaxillary, fully formed; fig. 7

is the displaced tooth which is cemented by the matrix to the palatal surface of the premaxillary in fig. 2. The enamelled crown shows the fine raised longitudinal ridges better developed than one usually sees them in modern Crocodiles. There are twenty-one alveoli on each side of the upper jaw.

In all the particulars in which the skull under description differs from that of the *Crocodilus toliapicus*, it departs further from the nilotic crocodile, and resembles more the Gavial-like Crocodile of Borneo; and as one of the old Egyptian names of the Crocodile, *Champsa*, has been applied generically to the Gavials by some recent Erpetologists, I have adopted the term ' *Champsoïdes*' to signify the resemblance of the present extinct species of Eocene Crocodile to the Gavials.

The basioccipital condyle, together with the condyloid processes of the exoccipital, project backwards in the *Croc. champsoïdes* farther than in any modern Crocodile; and the supraoccipital 3, fig. 4, T. III, descends nearer to the foramen magnum.

The upper jaw is more depressed, and the suborbital part of the maxillary bone is much less inclined to the vertical in the present skull than in the original of Dr. Buckland's figure of the *Crocodilus Spenceri*, which in other respects more nearly resembles the *Croc. champsoïdes* than the *Croc. toliapicus*; the difference above specified seems to be greater than can be accounted for by any accidental pressure to which the fossil skull figured in T. III can have been subjected. The mutilated skull to which the term *Croc. Spenceri* was originally applied, is defective, as I have said, in the facial or maxillary portion which is requisite for its unequivocal determination to either of the two species which the more perfect specimens since acquired have proved to have existed at the Eocene tertiary period. The form of the mutilated portion of skull, and the figure of it given in Pl. 25' of the ' Bridgewater Treatise,' might well appear to indicate a short and broad snouted species of true Crocodile; but if it be not distinct from the two better represented species above described, I should be more inclined to refer it to that which has the longest and narrowest snout, from the conformity of the characters of the part of the skull which is preserved. A view of the palatal surface of the specimen in question is given in T. II, fig. 2.

Crocodilian vertebræ referable to the two foregoing species of Sheppy Crocodiles.

Not more than two species of Crocodile are indicated by the detached vertebræ from Sheppy; but the different proportions of the homologous cervical vertebræ, fig. 3 and 7, T. V, and of the characteristic biconvex caudal vertebra, fig. 7, T. IV, and fig. 10, T. V, would have determined the fact of there being two distinct species, had their cranial characters, which are so satisfactorily demonstrated in T. II *A* and T. III, remained unknown. I refer, provisionally, the shorter and thicker vertebræ to the *Crocodilus toliapicus* with the shorter and thicker snout, and the longer and thinner vertebræ to the *Croc. champsoïdes* with the snout of similar proportions.

5

Vertebræ of the CROCODILUS TOLIAPICUS, Tab. IV and Tab. V, fig. 1, 2, 3, 5, 6.

The vertebra, fig. 1, 2, T. V, is the fourth cervical; it differs from that of the *Crocodilus acutus, Croc. Suchus,* and *Croc. biporcatus,* in the greater breadth and squareness of the base of the hypapophysis (fig. 2 *h*), which extends almost to the bases of the parapophyses *p*; the vertical diameter of the parapophyses is greater in comparison with their antero-posterior extent in the fossil than in the above-cited recent Crocodiles; the neurapophyses are thicker, and terminate in a more rounded border both before and behind; their bases extend inwards, and meet above the centrum, whilst a narrow groove divides them in the recent Crocodiles above cited; the length of the centrum is greater in proportion to the height and breadth in the fossil vertebra. In other respects the correspondence is very close, and the modern crocodilian characters are closely repeated. Traces of the suture between the centrum and neurapophysis remain, as shown at *n, n,* fig. 1. The diapophysis *d,* and the upper portion of the neural arch, with the zygapophyses and neural spine, have been broken away; the borders of the articular ends of the centrum have been worn away.

The vertebra (fig. 3, T. V) is the sixth cervical: in this specimen the base of the hypapophysis is contracted laterally and extended antero-posteriorly; the side of the centrum above the parapophysis (*p*) has become less concave; the vertebra has increased more in thickness than in length; in these changes it corresponds with the modern Crocodiles; it has been mutilated and worn in almost the same manner and degree as the fourth cervical.

The vertebra (fig. 1, 2, T. IV) is a seventh cervical of a smaller individual of the *Crocodilus toliapicus.* The hypapophysis has become more compressed and more extended antero-posteriorly; the parapophysis has become shortened antero-posteriorly, and increased in vertical diameter. The anterior concave surface of the centrum (fig. 1) is more circular, less extended transversely, than in the corresponding vertebra of the recent Crocodiles compared with the fossil.

Fig. 3, 4, T. IV, are two views of the eighth cervical of an individual of about the same size as that to which the fourth and sixth cervicals in T. V belong. Fig. 4, exemplifies the same difference which fig. 1 presents in regard to the more circular contour of the anterior concave surface of the centrum as compared with recent Crocodiles; the bases of the neurapophyses are thicker and more rounded anteriorly; the neural canal is rather more contracted; the base of the hypapophysis more extended in the axis of the vertebra (see fig. 3) than in the recent Crocodiles compared. The parapophyses have now risen, as in those Crocodiles, to the suture of the neurapophysis, and the diapophysis springs out at some distance above that suture.

Fig. 6, T. IV, shows the under surface of a dorsal vertebra, in which the hypapophysis ceases to be developed (probably the fourth or fifth).

Fig. 5, T. IV, gives the same view of one of the lumbar vertebræ, showing the elongation of the centrum, and the broad bases of the depressed diapophyses; there is an indication of two longitudinal risings towards the back part of the under surface of the centrum.

Fig. 5 and 6, T. V, give two views of the anterior sacral vertebra of the *Crocodilus toliapicus*; it is concave and much expanded transversely at its fore part (fig. 5), flattened and contracted behind. Traces of the suture remain to show the proportion of the anterior articular surface which is formed by the base of the pleurapophysis p; and fig. 6 shows the extension of that base from the side of the centrum upon the diapophysis or overhanging base of the neurapophysis; the under surface of the centrum of this vertebra has a slight median longitudinal rising.

Fig. 7, T. IV, gives a side view of the characteristic, biconvex, anterior caudal vertebra of the *Crocodilus toliapicus*.

Fig. 8, 9, T. IV, give two views of a middle caudal vertebra: in fig. 9 are shown the characteristic hypapophysial ridges extending from the articular surfaces for the hæmapophyses at the hind part of that aspect of the centrum: in fig. 8 the processes of the neural arch are restored in outline; a thick and low ridge extends from the middle of the side of the centrum to the base of the transverse process which it strengthens, like an underpropping buttress.

Vertebræ of the CROCODILUS CHAMPSOIDES.

Fig. 7 and 8, T. V, give two views of the third cervical vertebra of the above-named gavial-like Crocodile, which vertebra, besides its longer and more slender proportions, differs in the smaller size of its hypapophysis from the corresponding vertebra in any existing species of Crocodile or Gavial: the process in question being in the form of a low crescentic ridge, as shown at figure 8, between the bases of the parapophyses (p).

Both parapophyses terminate by a convex surface, which appears to have been a natural one. Between the parapophysis (p) and diapophysis (d), fig. 7, the side of the centrum is more deeply excavated than in the *Crocodilus toliapicus*. The centrum contributes a small part to the base of the diapophysis, as in the third cervical vertebra of modern Crocodiles. The neurapophysis are thinner than in the *Croc. toliapicus*, and their bases do not join one another above the centrum. The longitudinal ridge extending from the anterior to the posterior zygapophysis is sharply defined.

Fig. 4, T. V, is the first dorsal vertebra of the *Crocodilus champsoïdes*, in which, as in existing Crocodiles, the parapophysis (p) has passed almost wholly from the centrum upon the neurapophysis, the diapophysis (d) having been subject to a corresponding ascent. The base of the compressed hypapophysis extends over the anterior third of the middle line of the under surface of the centrum. There is a remarkable transverse constriction at the base of the posterior ball of the centrum, as if a string had been tied round that part when it was soft, and there is no appearance of this groove having been produced by any erosion of the fossil, or being otherwise than natural.

The same character is repeated, though with less force, in the posterior dorsal vertebra, fig. 9, T. V, and, together with the general proportions of the specimen, supports the reference of that vertebra to the *Crocodilus champsoïdes*. There is a slight longitudinal depression at the middle of the side of the centrum near the suture with the neurapophysis (*n, n*).

Fig. 10 is a side view of the first caudal vertebra of the *Crocodilus champsoïdes*: besides being longer and more slender than that vertebra is in the *Croc. toliapicus*, the inferior surface of the centrum is less concave from before backwards.

The evidences of Crocodilian reptiles from the deposits at Sheppy less characteristic of particular species than those above described, are abundant. Mr. Bowerbank possesses numerous rolled and fractured vertebræ, condyloid extremities, and other portions of long bones; with fragments of jaws and teeth.

Mr. J. Whickham Flower, F.G.S., has transmitted to me some fragments of the skull of a Crocodile from Sheppy, including the articular end of the tympanic bone, equalling in size that of a *Crocodilus biporcatus* the skull of which measures two feet eight inches in length.

Mr. Leifchild, C.E., possesses a considerable portion of the lower jaw of a Crocodile of at least equal dimensions, also from Sheppy, showing the angle of union of the rami of the lower jaw which corresponds with that in the *Crocodilus toliapicus*, Pl. 2.

In the museum of my esteemed and lamented friend, the late Frederic Dixon, Esq., F.G.S., at Worthing, is preserved a portion of the fossilized skeleton of a Crocodile, from the Eocene clay at Bognor, in Sussex; it consists of a chain of eight vertebræ, including the lumbar, sacral, and the biconvex first caudal, which are represented of their natural size in tab. xv, of Mr. Dixon's beautiful and valuable work on the 'Geology of Sussex.' A dorso-lateral bony scute adheres to the same mass of clay close to the vertebræ, and doubtless belonged to the same individual. The proportions of the vertebræ agree with those of the *Crocodilus toliapicus*. This fine specimen was discovered, and presented to Mr. Dixon, by the Rev. John Austin, M.A., Rector of Pulbrough, Sussex. Mr. Dixon had also obtained from the same locality a posterior cervical vertebra of a Crocodile, similar in its general characters to those above mentioned, but belonging to a larger individual. The length of the body of this vertebra is two inches and a half.

I have examined some remains of *Crocodilia* from the London Clay at Hackney; but as these also are not sufficiently perfect or characteristic for decided specific determination, no adequate advantage would be obtained by a particular description, or by figures of them.

The chief conclusion arrived at from the study of the Crocodilian fossils from the Island of Sheppy has been the proof, by the specimens selected for depiction in the present work, that at least two species of true *Crocodile* have left their remains in that locality; that neither of these had a short and broad snout like the Caimans, but that one of them—the *Croc. champsoïdes*—much more nearly resembled the Gavial of the Ganges in the proportion of that part of the skull; although, in its composition, especially as regards the length and connexions of the nasal bones, it is a true Crocodile.

Amongst the existing species of Crocodile the *Croc. acutus* of the West Indies offers the nearest approach to the *Croc. toliapicus*, and the *Croc. Schlegelii* of Borneo most resembles the *Croc. champsoïdes*. But there are well-marked characters in both the skull and the vertebræ which specifically distinguish the two fossil Crocodiles of Sheppy from their above-cited nearest existing congeners.

CROCODILUS HASTINGSIÆ, *Owen.* Tab. VI, VII, VIII, IX, and T. XII, fig. 2 and 5.
Reports of the British Association, 1847, p. 65.

That Crocodiles with proportions of the jaws assigned to the Eocene species noticed in Dr. Buckland's 'Bridgewater Treatise' and especially adapted for grappling with strong mammiferous animals, actually existed at that ancient tertiary epoch, and have left their remains in this island, is shown by the singularly perfect fossil skull figured in the above-cited plates. This specimen was discovered by the Marchioness of Hastings, in the Eocene fresh-water deposits of the Hordle Cliffs in Hampshire, which her Ladyship has described in the volume of 'Reports of the British Association' above cited, (p. 63).

When the specimen was originally exposed, it was in the same extremely fragile and crumbling state as the beautiful carapaces of *Trionyx* obtained by Lady Hastings from the same locality, and described and figured in the First Part of this Monograph on the *Chelonia;* but thanks to the skill and care with which the noble and accomplished discoverer readjusted and cemented the numerous detached fragments of those specimens, the present unique fossil has been in like manner restored as nearly to its original state as is represented in the plates; and all the requisite characters for determining the nature and affinities of the species, can now be studied with the same facility as in the skulls of existing Crocodiles.

If the reader will compare the plates above cited with the section of Cuvier's 'Ossemens Fossiles,' in which the distinctions between the Alligators and Crocodiles are specified,* he will see, (in fig. 1, T. VII) for example, that the fourth tooth or canine of the lower jaw is not received into a circumscribed cavity of the upper jaw,

---

* "Les têtes des caïmans, outre le nombre des dents, et surtout la manière dont la quatrième d'en bas est reçue, outre les différences qui dépendent de la circonscription totale, se distinguent de celles des Crocodiles proprement dits, 1°, parce que le frontal antérieur et le lacrymal descendent beaucoup moins sur le museau; 2°, en ce que les trous percés à la face supérieure du crâne, entre le frontal postérieur, le pariétal et le mastoïdien, y sont beaucoup plus petits, souvent même y disparaissent tout-à-fait, comme dans le caïman à paupières osseuses; 3°, en ce que l'on aperçoit une partie du vomer dans le palais, entre les intermaxillaires et les maxillaires; 4°, en ce que les palatins avancent plus dans ce même palais, et s'y élargissent en avant; 5°, en ce que les narines postérieures y sont plus larges que longues, etc." (tom. v, pt. ii, p. 105.)

but is applied to a groove upon the side of the upper jaw, and is exposed. Fig. 1, T. VI, shows that the prefrontal (14) and lachrymal (73) bones, instead of descending much less upon the facial part of the skull, extend much more, and advance nearer to the end of the muzzle than in any Alligator, or even than in any actual species of broad-nosed Crocodile.

The vacuities left between the postfrontal (12), the parietal (7), and the mastoid (8) (T. VI, fig. 1, and T. II, fig. 3), are as wide as in the skull of a *Crocodilus biporcatus* of equal size, and are larger than in the *Alligator lucius* or *All. sclerops*. Fig. 2, T. VII, shows that no part of the vomer is visible between the premaxillaries (22) and maxillaries (21), or elsewhere on the palate. But the palatine expansion of the vomer is not a constant character; it is wanting, for example, in the *Alligator lucius* of North America. The palatines (20) are not more advanced in the fossil in question than they are in the true Crocodiles, and their anterior portion does not expand to its anterior truncated termination. The posterior nostril, the entire contour of which is shown in the portion of the skull of the same species figured in T. VI, fig. 3, is longer than it is broad.

There is but one character in which the fossil skull in question differs from the true Crocodile, and agrees with most species of Alligator; it is in the reception of the two anterior teeth of the lower jaw into cavities of the premaxillaries, shown in fig. 2, T. VII, which are not perforated; so that there are no foramina anterior to the bony nostril, as in T. I, in the bone marked 22. These foramina are not, however, absent in all Alligators; the skull of the *Alligator sclerops*, figured by Cuvier (tom. cit., pl. i, fig. 7), shows them, as do all the species of true Crocodile the skulls of which are figured in the same plate. There is one character by which the *Crocodilus Hastingsiæ* differs from all known species of both Crocodile and Alligator: it is that afforded by the broad and short nasal bones (15, fig. 1, T. VI), which do not reach the external nostril; this being formed, as in the Gavials, exclusively by the premaxillaries 22.

In the general proportions, however, of the skull of *Croc. Hastingsiæ*, especially the great breadth, shortness, and flatness of the obtusely-rounded snout, it resembles that of the Alligators more than that of any known species of true Crocodile, the length from the tympanic condyle to the end of the snout being to the breadth taken at the condyles as 16 to 9.

The following are dimensions of the fossil in question:

	Feet.	Inches.	Lines.
Length of skull from the angle of the lower jaw to the end of the snout	1	6	6
Do. from the tympanic condyle to ditto.	1	4	6
Do. do. to the orbit	0	5	4
Do. from the orbit to the external nostril	0	7	0
Breadth of the skull across the tympanic condyles	0	9	3
Do. the orbits	0	7	0
Do. the external nostril	0	4	0
Longest diameter of upper temporal aperture	0	1	9
Do. the post-palatal vacuities	0	4	9
Depth of the lower jaw at the posterior vacuity	0	3	0
Depth of the occipital region	0	4	3

The occipital region of the skull (T. VI, fig. 2), in the proportion of its breadth to the depth of the lateral parts formed by the conjoined paroccipitals (4) and mastoids (8), resembles that of the true Crocodiles rather than that of the Alligators, in which that region is proportionally deeper than in the Crocodiles; the vertical extent of the supraoccipital is less, and that of the conjoined parts of the exoccipitals above the foramen magnum is greater; the vertical extent of the descending part of the basioccipital is also greater in proportion to its breadth than in the Alligators. The proportion of the basisphenoid (5) and of the conjoined parts of the pterygoids (24) which appear in this view (fig. 2), is less than in the Alligators, but is greater than in most Crocodiles, thus presenting an intermediate character; but the entire exclusion of any part of the posterior nostril from this view is a character of the Alligators, and is due to the horizontal plane of that aperture in them, and to its position in advance of the posterior border of the pterygoids, from which it is partitioned off usually by a bony ridge. The posterior nostril has the same position and aspect in the *Crocodilus Hastingsiæ*, and these characters of the posterior nostril are perhaps more distinctive between Alligator and Crocodile than the shape of the aperture, in which the present fossil differs both from the Alligators and from most of the Crocodiles with which I have compared it. The backward extension of the exoccipitals and of the basioccipital condyle, is such as to bring both parts into view in looking directly upon the middle of the upper surface of the skull, as in T .VI, fig. 1. In this character the fossil resembles the Crocodiles more than the Alligators, but the projection is greater than in existing Crocodiles, and equals that in the Sheppy *Crocodilus champsoïdes*.

On the upper surface of the skull a distinctive character has been pointed out by Cuvier in the different proportions of the supra-temporal apertures in the Alligators and Crocodiles. The horizontal platform in which these apertures are perforated, is also square in the Alligators; the mastoidal angles being less produced outwards and backwards, and the postfrontal angles less rounded off; this difference is shown in the skulls figured in Cuvier's pl. i, tom. cit. The *Croc. Hastingsiæ*, both by the obtuseness of the postfrontal angles, and the acuteness and production of the mastoidal angles, resembles the Crocodiles, as well as by the size of the supra-temporal apertures; these are ovate with the small end turned forwards and a little outwards.

Another character may be noticed in the figures of the skulls of the three species of Alligators as compared with those of the three species of Crocodile in Cuvier's pl. i, viz. the larger proportional size of the orbits in the former, in which the orbit much exceeds in size the lateral temporal aperture. In the *Alligator niger*, also, I find the orbits enormous, and it is the encroachment of the narrow anterior part of the orbital cavity upon the facial part of the prefrontal and lachrymal, that renders that part of those bones relatively shorter in the Alligators. In the *Crocodilus Hastingsiæ* the proportions of the lateral temporal apertures (T. VI, fig. 1, 12, 26) and orbital (11, 14, 73) apertures, are like those in the species of Crocodile in which the orbits are smallest. The extent of the facial part of the prefrontal (14) and lachrymal (73) is greater in the *Croc. Hastingsiæ* than in any existing species of true Crocodile. Another characteristic of the present fossil presented by the upper surface of the skull, is the shortness as well as breadth of the nasal bones, and their almost truncate anterior termination at nearly one inch from the external nostril. In all the Alligators' skulls that I have examined or seen figured, the nasal bones are broadest at their posterior third part, and converge to a point anteriorly, where in the *Alligator lucius*, e. g., they extend across the nasal aperture.

The interorbital space is slightly concave in the *Crocodilus Hastingsiæ*; two broad and slightly elevated longitudinal tracts are continued forwards upon the face from the fore part of the orbits; but they are not developed into ridges, as in the *Croc. biporcatus*. The maxillaries swell out a little in advance of the middle of the nasals, and then contract to the crocodilian constriction at the suture with the premaxillaries, where the tips of the lower canines appear in the upper view (fig. 1, T. VI), and their whole crown is exposed in the side view (fig. 1, T. VII). The conjoined parts of the premaxillaries send a short pointed projection into the back part of the external nostril.

On the under or palatal surface of the skull (T. VII, fig. 2) the maxillo-premaxillary suture runs almost transversely across, as in the *Crocodilus rhombifer*, figured by Cuvier in pl. iii, fig. 2, of the volume above cited. There is no appearance of the vomer upon the palate. The palatal bones (20), though somewhat broader anteriorly, and more abruptly truncate than in any existing Crocodile that I have seen, are more like those bones in the true Crocodiles than in the Alligators. The portion between the post-palatal vacuities is longer and narrower; the posterior end of the palatines is narrower, and the part of the bone anterior to the notch receiving the posterior angle of the palatal plate of the maxillary does not expand in advancing forwards, as it does in the Alligators: in the *Alligator niger* this expansion is greater than in the *All. lucius*, and the posterior ends of the palatines are also remarkably expanded, and applied to the anterior borders of the pterygoids almost as far as their articulation with the ectopterygoids, the postpalatal vacuities not at all encroaching on the pterygoids, as they are seen to do at 24, T. VII, fig. 2, and also in the figure of the *Crocodilus rhombifer* above cited, and in other true Crocodiles. The form of the pterygoids (24, T. VII, fig. 2) is peculiar in the *Crocodilus Hastingsiæ*: they are contracted anteriorly, and send forwards a short truncated process to meet the narrow posterior ends of the palatines (20); and the same character being repeated in another skull of the same species, from Hordle, also in the collection of Lady Hastings, in which this part of the bony palate (T. VI, fig. 3) is more perfect than in the subject of T. VII, fig. 2, it may be regarded with some confidence as specific. In the *Crocodilus champsoïdes* of Sheppy it will be seen, by fig. 2, T. II, that the pterygoids (24, 24) are not produced where they join the palatines (20). In the Alligators, the posterior border of the conjoined pterygoids is deeply notched behind the posterior nostrils, the angles of the notch being slightly extended backwards: in most Crocodiles, the sides of the notch are so developed that it does not sink deeper than the line·of the posterior border of the pterygoids; and this modification is exaggerated in the *Crocodilus Hastingsiæ* (T. VI, fig. 3) in which the notch in question is merely the interval between two slender diverging processes from the middle of the back part of the pterygoids, 24. The posterior aperture of the nasal passages is wholly surrounded in the *Crocodilus Hastingsiæ* by the horizontal plate of the pterygoids, and has the same position and aspect as in the Alligators; but its form is heart-shaped, with the apex directed backwards, and the antero-posterior diameter exceeding the transverse one. I have not met with this form of the posterior nostril in any other species of Crocodilian; but it is repeated in two individuals of the *Croc. Hastingsiæ*, and may be regarded as a specific character.

The ectopterygoid, 25, T. VI, fig. 3, T. II, fig. 2 (*d*, fig. 2, pl. iii, 'Ossemens Fossiles,' t. v, pt. ii) articulates with a larger proportion of the outer surface of the pterygoids (24) in the Crocodiles than in the Alligators: it agrees with the Crocodiles in the extent of this articulation in the *Croc. Hastingsiæ*.

The number of teeth in this species is $\frac{22-22}{20-20} = 84$.

In the upper jaw the fourth, ninth, and tenth are the largest; and the fifteenth and sixteenth exceed in size those immediately before and behind them. The alveolar border of the jaw increases in depth to form the sockets requisite for firmly lodging these larger teeth, and gives rise to the festooned outline of the jaw, which is found in all Crocodiles and Alligators in proportion as the teeth are unequal in size.

The lower jaw presents the same compound structure as that in the *Crocodilia*, with the general form characteristic of that in the Alligators and in most of the true Crocodiles: the symphysis, e. g. is as short as *Crocodilus biporcatus* and the *Alligator niger*, in which it extends as far back as the interval between the fourth and fifth socket. This is the relative position of the back end of the symphysis in a fine and perfect under jaw of the *Crocodilus Hastingsiæ* in the collection of the Marchioness of Hastings. In a portion of the under jaw of apparently the same species of Crocodile, from the same locality, in the collection of Searles Wood, Esq., F. G. S., the symphysis terminates opposite the interval between the third and fourth tooth.

The chief distinction observable between the modern Crocodiles and Alligators in the lower jaw is the greater relative size of the vacuity between the angular (30) and surangular (29) pieces, and the greater relative depth of the ramus at that part, in the Alligators. In these characters the lower jaw of the present species more resembles that of the true Crocodiles; although, as the vacuity in question is somewhat larger, a slight affinity to the Alligator might be inferred from that circumstance. The comparative figures of the hinder third of the mandibular ramus in T. XII, fig. 4, 5, 6, will exemplify the difference in question, and the degree of proximity to the crocodilian and alligatorial characters respectively.

With regard to another character deducible from the relation of the backwardly-produced angle of the jaw to the articular surface, the *Crocodilus Hastingsiæ* more decidedly resembles the Alligator: I allude to the depth of the excavation between the articular cavity (29) and the end of the angular bone (30), and to the lower or higher level of the angle itself: the fossil jaw (fig. 5) resembles the Alligator (fig. 6) in this respect more than the Crocodile (fig. 4).

The alveoli are twenty in number in each ramus of the *Crocodilus Hastingsiæ*: the third and fourth are large, of equal size, and close together; behind these the eleventh, twelfth, and thirteenth are the largest, and the alveolar ridge is raised to support them; after the seventeenth the summits of the crowns of the teeth become obtuse, and the crowns mammilloid, and divided by a constriction or neck from the fang; they each, however, have a separate socket, as in the Crocodiles, the septa not being incomplete at the hinder termination of the dental series, as in the *Alligator niger* figured in my 'Odontography.'*

Fig. 3, T. II, gives a representation, of the natural size, of the cranial platform of a young *Crocodilus Hastingsiæ* in the collection of Searles Wood, Esq.; the hemispheric depressions in the surface of the bone are more regular, distinct, and relatively larger, and the interorbital part of the frontal is narrower, concomitantly with the larger proportional eyeballs and orbits of the young animal. The relatively larger supratemporal apertures form another character of nonage; but there is no ground for deducing a specific distinction from any of the differences observable between this part of the young crocodile's cranium and the corresponding part of that of the more mature specimen (T. VI).

ALLIGATOR HANTONIENSIS, *Wood.* Tab. VIII, fig. 2.

London Journal of Palæontology and Geology.

On reviewing the characters of the skull of the *Crocodilus Hastingsiæ* we perceive that they combine to a certain extent those which have been attributed to the genus *Crocodilus* and the genus *Alligator*; in general form it resembles most the latter, but agrees with the former in some of the particulars that have been regarded by Cuvier and other palæontologists as characteristic of the true Crocodiles. I allude more particularly to the exposed position of the inferior canines when the mouth is shut. Respecting which, however, I am disposed to ask, whether this be truly a distinctive character of importance? One sees that it needs but a slight extension of ossification from the outer border of the groove to convert it into a pit; yet the character has never been found to fail as discriminative of the several species of existing Crocodiles and Alligators hitherto determined. It constitutes, however, the only difference between the skulls of the *Crocodilus Hastingsiæ* in the collection of the Marchioness of Hastings and that fine portion of skull now, by the kindness of Mr. Searles Wood, before me, on which he has founded the species named at the head of the present section. So closely, in fact,

* Tom. ii, pl. lxxv, fig. 3.

do those specimens from the same rich locality correspond, that any other comparative view than that given in T. VIII appeared superfluous. In both the broad nasal bones terminate at the same distance from the external nostril, which is accordingly formed exclusively by the premaxillaries; in both, the palate-bones present the same narrow, truncate posterior ends, and the same equal breadth of their anterior portions included between the maxillaries; only these terminate rather more obliquely in Mr. Wood's specimen, their anterior ends forming together a very obtuse angle directed forwards. But this is comparatively an unimportant difference, and I regard as equally insignificant the slight interruption of the transverse line of the maxillo-premaxillary suture, at the middle part, which will be seen by comparing fig. 2 with fig. 1, in T. VIII. The teeth are the same in number, arrangement, and proportion in the *Alligator Hantoniensis* as in the *Crocodilus Hastingsiæ*, and the alveolar border of the jaws describes the same sinuous course.

Had the complete fossil skull first submitted to my inspection at the meeting of the British Association at Oxford presented the same fossæ for the reception of the lower canines which exist in fig. 2, T. VIII, I should have referred it to the Alligators, notwithstanding the crocodilian characters of the small orbits, the long facial plates of the prefrontal and lachrymal, the wide supratemporal apertures, the non-expansion of the fore part of the palatines, and the non-appearance of the vomer on the palate, with other minor marks of the like affinity. For all these characters arise out of secondary modifications, and are presented in different degrees in the different species of *Crocodile*, and are rather of a specific than a generic value. They determine the judgment by the extent of their concurrence rather than by their individual intrinsic worth, and for that reason, therefore, the exposed position of the lower canine in the lateral groove of the upper jaw inclined the balance in favour of a reference of the previously-described fossil to the true Crocodiles. One cannot, indeed, attach any real generic importance to the modification of the upper jaw in relation to the lower canines. In three examples, however, in the collection of the Marchioness of Hastings, the crocodilian modification of this character is repeated, as it is shown in T. VII, fig. 1; and we have to choose, therefore, between the conclusion that Mr. Wood's specimen (T. VIII, fig. 2) presents an accidental variety in this respect, or to view the fossæ in the upper jaw as indicative of not only a different species but a distinct genus from the *Crocodilus Hastingsiæ*. I should be glad to have more evidence on this point, and especially the opportunity of comparing the posterior nostrils, the orbits, the supra-temporal apertures, and the occipital part of the skull of a specimen from Hordwell, repeating the alligatorial character of the fossæ in the upper jaw for the lower canines. I am disposed to regard this character, notwithstanding its constancy in the living species of Alligator, as a mere variety in the Hordwell fossil; but pending the acquisition of further evidence, it seems best to record this fossil under the title proposed for it by the able geologist by whom it was discovered.

CROCODILUS HASTINGSIÆ.

Vertebræ referable to the CROCODILUS HASTINGSIÆ, *Tab. IX.*

The fossil crocodilian vertebræ obtained from the Eocene sand at Hordle, notwithstanding the comparatively limited extent of the researches in that interesting formation, are at least as abundant as those which have been discovered at Sheppy, but they do not, as at that locality, indicate two distinct species; all that have, hitherto, been found belong to one and the same kind of Crocodile, and from their robust proportions, would seem to have come from a species with a short and broad muzzle, like that of the Crocodile or Alligator, the fossil skulls of which have been described.

Perhaps the most perfect fossil reptilian vertebra that has hitherto been discovered is the one figured, of the natural size, in T. IX, fig. 1, 2, and 3. It is the fifth cervical vertebra. As compared with that of the *Crocodilus toliapicus* (T. V, fig. 1, 2), which it resembles in size, the hypapophysis, *hy* (fig. 2, T. IX), is much more compressed, and the under part of the centrum is more extensively and deeply excavated between it and the parapophyses (*p*); it is also excavated on each side behind the base of the hypapophysis, from which a progressively widening smooth ridge is continued to near the posterior surface of the centrum. The interspace at the side of the vertebra, between the parapophysis and diapophysis, is smaller but deeper in the *Crocodilus Hastingsiæ.* The neurapophyses meet above the centrum in both; but in the *Crocodilus Hastingsiæ* they are thicker anteriorly and thinner at their posterior border, and the neural canal (fig. 2, *n*) is more contracted than in the *Crocodilus toliapicus.*

As compared with the cervical vertebra of the *Crocodilus champsoïdes* from Sheppy, the present vertebra differs in the form of the hypapophysis in a greater degree than from the *Crocodilus toliapicus.* Fig. 8, T. V, shows as little as does fig. 2 in the same plate, the median ridge and lateral excavations of the under part of the centrum which characterise the present vertebra of the *Crocodilus Hastingsiæ.* The *Crocodilus champsoïdes* resembles the *Crocodilus Hastingsiæ* in the character of the proportion and depression of that part of the side of the centrum forming the interspace between the parapophysis and diapophysis; but the antero-posterior extent of the parapophysis is relatively less in that Sheppy species. The outer surfaces of the neurapophyses in the *Crocodilus Hastingsiæ* slope or converge towards each other from before backwards, in a much greater degree than in either of the Sheppy species. I have not observed in any recent Crocodile or Alligator the median ridge, continued backwards from the hypapophysis and the lateral depressions, so strongly developed, as in the *Crocodilus Hastingsiæ.* The fore part of the neurapophyses is relatively thicker in this than in the recent species. The pleurapophyses *pl*, (figs. 1, 2), are well developed both forwards and backwards, and the latter productions are expanded and excavated above for the reception of the fore part of the succeeding cervical rib. The zygapophyses (*z*) are thicker at their base, especially the hinder pair, where the base fills up the entire interval between the articular surface and the base of the spine (see fig. 2). There is the usual deep excavation at the fore and back part of the base of the spine (*ns*) for the insertion of the interspinal ligaments. The neural spine is compressed, moderately long, straight and truncate at its summit.

Although the hypapophysis maintains its characteristic form with much constancy in the homologous vertebræ of the same species of Crocodile, it varies in different cervical vertebræ of the same individual in certain existing species. It is, for example, shorter and thicker in the third and fourth vertebræ than in the succeeding ones in the *Crocodilus acutus*; whilst in the *Crocodilus biporcatus* the hypapophysis of the third cervical is more compressed than that of the sixth. The greatest difference is, however, presented, as far as I have yet made the comparison, by the cervical vertebræ of the *Alligator lucius*, in respect of the hypapophysis, which is broad and short in the third and fourth cervicals, but becomes long and slender in the succeeding cervicals. The small vertebral centrum (fig. 4, T. IX) resembles, in its broad and stunted hypapophysis, that of the third cervical vertebra of the Alligator, but with an indication of a median rising and lateral depressions, behind that process, like those which are more decisively shown in the fifth cervical vertebra of the larger individual of the *Crocodilus Hastingsiæ*, to which species I believe the specimen fig. 4 to belong. It is the homologous vertebra with fig. 8, T. V, and well illustrates the different proportions of the bones in different species of Crocodile.

Fig. 6 gives a view of the anterior surface of the first sacral vertebra of the *Crocodilus Hastingsiæ*: the under surface of the centrum has ceased to develope the median ridge; the short and thick ribs (*pl*) have completely coalesced with both the centrum and neural arch. The anterior concavity has a fuller and more exact elliptical form than that of the *Crocodilus toliapicus* (fig. 5, T. V); the anterior zygapophyses do not project over the rim of that concavity; but, like those of the Alligator and Crocodile, they are more transversely extended than in the Gavial.

The general proportions of the first caudal vertebra (fig. 7, T. IX) are intermediate between those of the *Crocodilus toliapicus* (fig. 7, T. IV) and of the *Crocodilus champsoïdes* (fig. 10, T. V): the under surface of the centrum is flat, not concave, lengthwise, as in both the Sheppy Crocodiles; the side of the centrum is irregularly tuberculate, not smooth, and concave lengthwise; the broad and high neural spine is deeply grooved at its fore part: a smaller proportion of the hinder end of the centrum (fig. 5) is occupied by the articular ball than we find in the antecedent vertebræ.

As none of the other numerous vertebræ and portions of vertebræ give any indications of a different species from the *Crocodilus Hastingsiæ*, or add any material characters to those of that species which have been deduced from the parts of the skeleton already described, I refrain from trespassing on the reader's attention or occupying further space by their description or figures.

Genus—GAVIALIS, Oppel.

GAVIALIS DIXONI, *Owen.*　　Tab. X.

The characters of the genus *Gavialis* are much more strongly marked than are those which distinguish the Alligators from the Crocodiles, and leave no ambiguity in the conclusions that may be deduced from them.　The present interesting addition to the catalogue of British Fossil Reptiles, is due to the discovery in the Eocene deposits at Bracklesham, by my lamented friend the late Frederic Dixon, Esq., F.G.S., of the remains figured in T. X.　The portions of the lower jaw demonstrate, by the slender proportions of the mandibular rami (figs. 1, 5), the extent of the symphysis, the uniform level of the alveolar series, and the nearly equal distance of the sockets of the comparatively small, slender, and equal-sized teeth, the former existence in England, during the early tertiary periods, of a Crocodilian with the maxillary and dental characters of the genus *Gavialis*.　These characters are, however, participated in by some of the extinct Crocodilians of the secondary strata (see T. XI, fig. 2′); but in them they coexist with a different type of vertebra from that of the recent and known tertiary Crocodilian genera : it became necessary, therefore, to ascertain what form of vertebra might be so associated with the fossil Gavial-like jaws and teeth in the Bracklesham Eocene deposits, as to justify the conclusion that such vertebræ had belonged to the same species as the jaws.　Now, the only Crocodilian vertebræ that have yet been found at Bracklesham, so far as I can ascertain, present the procœlian type of articular surfaces of the body (T. X), like that in Mr. Dixon's collection fig. 8.　This vertebra answers to the fifth cervical vertebra in the existing Crocodilians, and accords in its proportions with that in the Gangetic Gavial.　There are a few indications of specific distinction; the parapophysis (p) or lower transverse process articulating with the head of the rib, is relatively shorter antero-posteriorly.　The broad, rough, neurapophysial sutures (n) meet upon the middle of the upper part of the centrum; the elsewhere intervening narrow neural tract sinks deeper into the centrum than in the modern Gavial, but is perforated, as in that species, by the two approximated vertical vascular fissures.　The hypapophysis (hs) or process from the inferior surface of the centrum, has been broken off in the fossil, but it accords in its place and extent of origin with that in the fifth and following cervical vertebræ of the Gavial.　Assuming the fossil procœlian vertebræ from Bracklesham, and the above-described vertebra in particular, to have belonged to the same individual or species as the portions of fossil jaw (figs. 1, 5), then these mandibular and dental fossils must be referred to the genus *Gavialis*, or to the long-, slender-, and subcylindrical-snouted *Crocodilia* with procœlian vertebræ.

This genus is now represented by one or two species peculiar to the great rivers of India, more especially the Ganges; and the fossil differs from both the *Gavialis gangeticus*, Auct., and from the (perhaps nominal) *Gavialis tenuirostris*, Cuv., in the form and relative size of the teeth. The crown (figs. 6, 7) is less slender in the fossil than in the existing Gavials, and less compressed, its transverse section being nearly circular. There are two opposite principal ridges, but they are less marked than in the existing Gavials; and are placed more obliquely to the axis of the jaw, i. e., the internal ridge is more forward, and the external one more backward, when the tooth is in its place in the jaw. In the modern Gavial, the opposite ridges, besides being more trenchant, are nearly in the same transverse line. The other longitudinal ridges on the enamel of the fossil teeth, are more numerous, more prominent, and better defined, than in the existing Gavials: the intermediate tracts of enamel present the same fine wrinkles in the fossil as in the existing Gavials' teeth.

The two chief portions of jaw (fig. 1, and figs. 4, 5) belong to two individuals of different ages; indicated by the difference in the breadth and depth of the ramus: both specimens being from the corresponding part of the jaw, viz. where it forms the long symphysis characteristic of the Gavials. The specimen (figs. 4, 5) includes a larger proportion of the jaw than the fragment delineated in fig. 1.

On comparing the latter fragment of the fossil lower jaw with a specimen of a lower jaw of the *Gavialis gangeticus* of the same breadth across the symphysial part, at the intervals of the sockets, which breadth is 3 centimeters (1 inch 3 lines), I find that the longitudinal extent of 10 centimeters (near 4 inches) of a ramus of the fossil jaw includes five sockets; but in the recent Gavial the same extent of jaw includes seven sockets, showing that the teeth are fewer as well as larger in the fossil Gavial, in proportion to the breadth of the jaws.

The second portion of the jaw (fig. 2) is from the part where the rami diverge posteriorly from the symphysis, and near the posterior termination of the dentary series. Here the teeth become shorter in proportion to their thickness, and somewhat closer placed together: there is a shallow depression (*c*) in each interspace of the teeth, for the reception of the crowns of the opposite teeth when the mouth is shut. These depressions are longer, deeper, and better defined in the fossil than in the recent Gavial of the same size.

The fragments of jaw and teeth of the fossil Gavial of Bracklesham show examples of young teeth penetrating the base of the old ones, according to the law of succession and shedding of the teeth, which characterises the existing *Crocodilia*: fig. 2 shows the apex of one of the successional teeth at *d*; and fig. 3 *d* the hollow base of the same incompletely formed tooth seen from below.

Besides the fossil jaws, teeth, and vertebræ of the extinct Gavial, a nearly entire femur (fig. 9) of a Crocodilian has been discovered in the Eocene deposits at Bracklesham, which in its proportions, agrees with that bone in the Gavial of the Ganges. Cuvier, in his comparison of the bones of the Gavial with those of the Alligators and true Crocodiles, merely observes, " La forme des os du Gavial ressemble aussi prodigieusement à celle des os du Crocodile, seulement les apophyses épineuses des vertèbres sont plus carrées."*

With regard to the femur, this bone is more slender in proportion to its length in the Gangetic Gavial, than in the *Crocodilus biporcatus* or the *Alligator lucius,* and the anterior convex bend of the shaft commences nearer the head of the bone; and in these characters the fossil femur from Bracklesham corresponds with the modern Gavial, and differs from the Crocodiles and Alligators, and also from the *Crocodilus Hastingsiæ,* of which species specimens of the fossil femur have been kindly submitted to me by the Marchioness of Hastings and Alexander Pytts Falconer, Esq. The fossil femur of the Gavial from Bracklesham (fig. 9) may therefore be referred, with the utmost probability, to the same species as the portions of jaw, teeth, and vertebræ above described; and as these clearly demonstrate a species distinct from any known Gavial, I propose to call the extinct species of the Eocene deposits at Bracklesham, *Gavialis Dixoni,* after my esteemed friend, by whose scientific and zealous investigations so much valuable additional knowledge has been obtained respecting the fossils of that rich, but, previously to his researches, little known locality.

The tooth represented of the natural size in fig. 10, T. X, was also discovered at Bracklesham, and forms part of the collection of G. Coombe, Esq. It resembles, in its proportions and obtuse extremity, the teeth of the Crocodiles rather than those of the Gavials, and at first sight reminded me of those of the *Goniopholis* or amphicœlian Crocodile of the Wealden period. On comparing it closely with similar-sized teeth of that species, the enamel ridges were more numerous and decided in the *Goniopholis;* and the delicate reticular surface in the interspaces of the more widely separated and feebler longitudinal ridges in the Bracklesham tooth was wanting in the *Goniopholis.* The minute superficial characters of the enamel of the large and strong Crocodilian tooth from Bracklesham, closely agree with those of the *Gavialis Dixoni.* It is just possible that this may be a posterior tooth of a very large individual of that Gavial, as the teeth become at that part of the jaw shorter in proportion to their thickness in the modern Gavials. If it should not belong to that Gavial, it must be referred to a Crocodile distinct from those species of the secondary strata, or those existing Crocodiles which have teeth of a similar form; since they present a different superficial pattern of markings on the enamel.

On reviewing the information which we have derived from the study of the fossil remains of the procœlian *Crocodilia,* that have been discovered in the Eocene deposits of England, the great degree of climatal and geographical change, which this part of Europe must have undergone since the period when every known generic form of that group of reptiles flourished here, must be forcibly impressed upon the mind.

At the present day the conditions of earth, air, water, and warmth, which are

* Ossemens Fossiles, 4to, tom. v, pt. ii, p. 108.

indispensable to the existence and propagation of these most gigantic of living Saurians, concur only in the tropical or warmer temperate latitudes of the globe. Crocodiles, Gavials, and Alligators now require, in order to put forth in full vigour the powers of their cold-blooded constitution, the stimulus of a large amount of solar heat, with ample verge of watery space for the evolutions which they practise in the capture and disposal of their prey. Marshes with lakes, extensive estuaries, large rivers, such as the Gambia and Niger that traverse the pestilential tracts of Africa, or those that inundate the country through which they run, either periodically, as the Nile for example, or with less regularity, like the Ganges ; or which bear a broader current of tepid water along boundless forests and savannahs, like those ploughed in ever-varying channels by the force of the mighty Amazon or Oronooko ;—such form the theatres of the destructive existence of the carnivorous and predacious Crocodilian reptiles. And what, then, must have been the extent and configuration of the eocene continent which was drained by the rivers that deposited the masses of clay and sand, accumulated in some parts of the London and Hampshire basins to the height of one thousand feet, and forming the graveyard of countless Crocodiles and Gavials ? Whither trended that great stream, once the haunt of Alligators and the resort of tapir-like quadrupeds, the sandy bed of which is now exposed on the upheaved face of Hordwell Cliff ?

Had any of the human kind existed and traversed the land where now the base of Britain rises from the ocean, he might have witnessed the Gavial cleaving the waters of its native river with the velocity of an arrow, and ever and anon rearing its long and slender snout above the waves, and making the banks re-echo with the loud and sharp snappings of its formidably-armed jaws. He might have watched the deadly struggle between the Crocodile and Palæothere, and have been himself warned by the hoarse and deep bellowings of the Alligator from the dangerous vicinity of its retreat. Our fossil evidences supply us with ample materials for this most strange picture of the animal life of ancient Britain, and what adds to the singularity and interest of the restored ' tableau vivant,' is the fact that it could not now be presented in any part of the world. The same forms of Crocodilian Reptile, it is true, still exist, but the habitats of the Gavial and the Alligator are wide asunder, thousands of miles of land and ocean intervening : one is peculiar to the tropical rivers of continental Asia, the other is restricted to the warmer latitudes of North and South America ; both forms are excluded from Africa, in the rivers of which continent true Crocodiles alone are found. Not one representative of the Crocodilian order naturally exists in any part of Europe ; yet every form of the order once flourished in close proximity to each other in a territory which now forms part of England.

Order—LACERTILIA.

PLEURODONT LIZARD (?)　Tab. XIV, figs. 43, 44.

Although members of the present order, with the modern procœlian type of vertebræ, existed in England during the Wealden and Chalk periods, and the greater part of the actual class of Reptiles, in all parts of the world, is composed of the same order, yet but one solitary example of true Lacertian from the formations of the Eocene tertiary period has hitherto come under my observation—a fact which has often excited my surprise.　Future researches may bring to light farther and better evidence of the class.

Among the fossils obtained by Mr. Colchester from the Eocene sand, underlying the Red Crag at Kyson, or Kingston, in Suffolk, the existence of a Lizard, about the size of the Iguana, is indicated by a part of a lower jaw, armed with close-set, slender, subcylindrical, antero-posteriorly compressed teeth, attached to shallow alveoli, and with their bases protected by an external parapet of bone.　The fragment of jaw is traversed by a longitudinal groove on the inside (fig. 44), and is perforated, as in most modern Lizards, and as in some Fishes, by numerous vascular foramina along the outside (fig. 43).　The teeth are hollow at their base.

TAB. I.

Crocodilus Suchus, half the nat. size.

Fig.

1. Upper view of the skull.

2. Under view of ditto.   Both ends of the bony palate have suffered fracture and loss, apparently in the process of mummifying the body of the Crocodile.

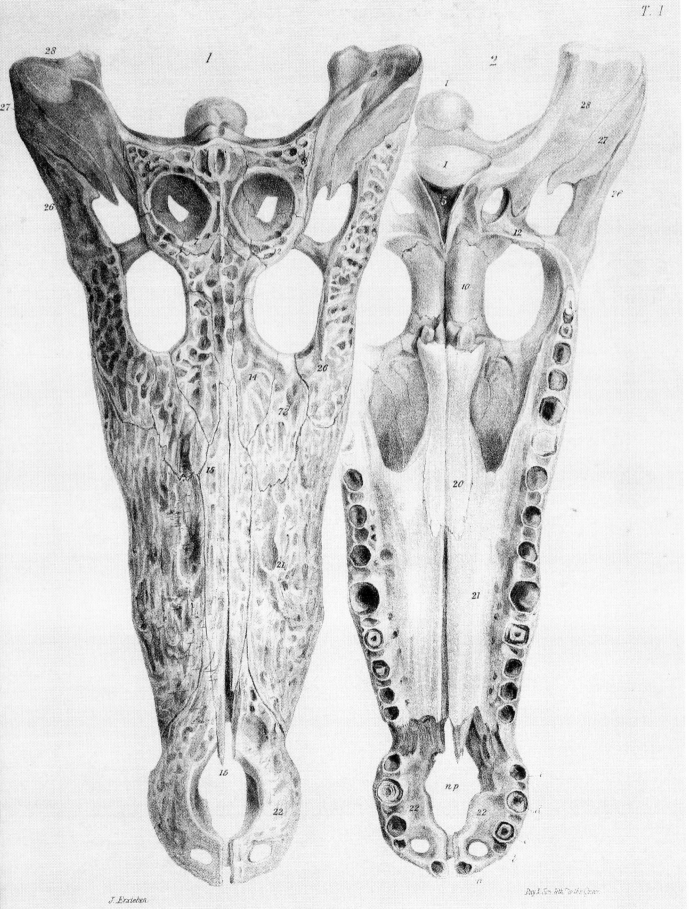

Crocodilus Suchus. *(From an Egyptian Mummy.)*

TAB. II.

Fig.

1. Upper view of the skull of the *Crocodilus toliapicus*, one third the nat. size.

2. Under view of the hinder portion of the skull of the *Crocodilus champsoïdes*, on which Dr. Buckland founded his species of 'Crocodile with a short and broad snout,' called *Crocodilus Spenceri*; half the nat. size.

3. Upper view of the cranial platform of a young *Crocodilus Hastingsiæ*, nat. size.

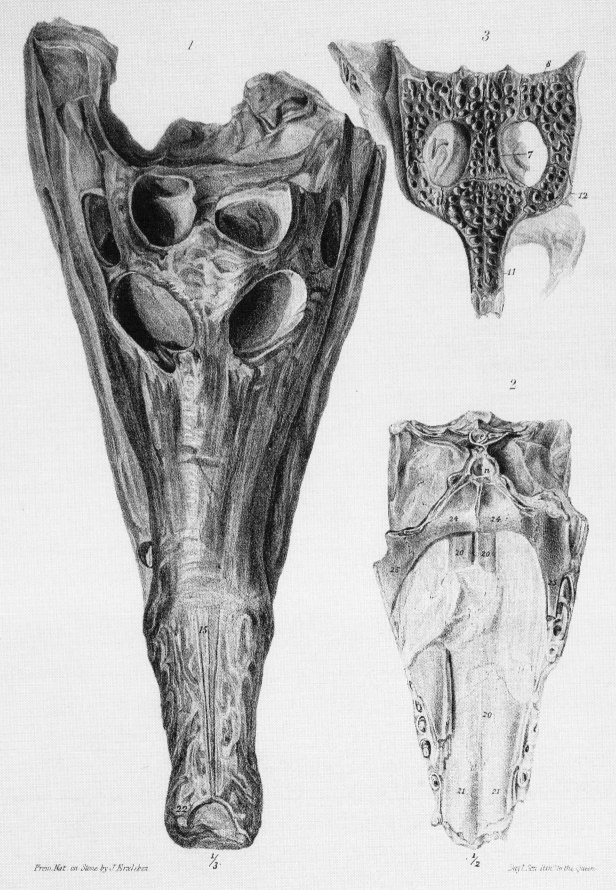

1

3

8

7

12

11

2

n

24   24

20   20

25

20

21   21

22

15

From Nat. on Stone by J.Erxleben   1/3   Day & Son, Lith to the Queen.

1/2

TAB. II *A.*

Crocodilus toliapicus, one third the nat. size.

Fig.
1. Oblique side view of the skull.

2. Under view of the skull.

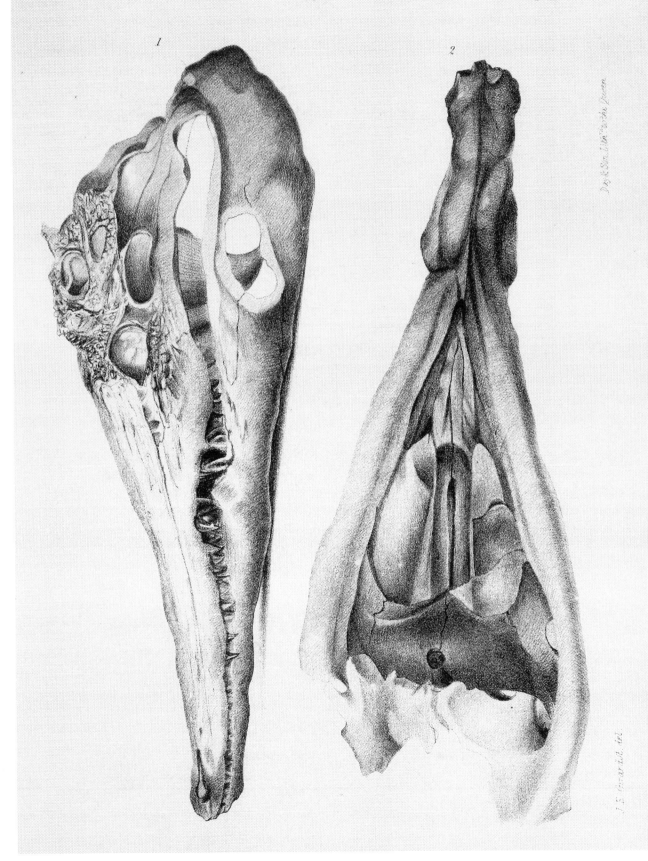

TAB. III.

Crocodilus champsoïdes, half the nat. size.

Fig.
1. Upper view of the skull.
2. Under view of the fore part of ditto.
3. Side view of ditto.
4. The left half of the back part of the skull of ditto, nat. size.
5. The point of an anterior young tooth coming into place, nat. size.
6. Half of the crown of one of the large conical teeth, nat. size.
7. One of the shorter and more obtuse posterior teeth, nat. size.

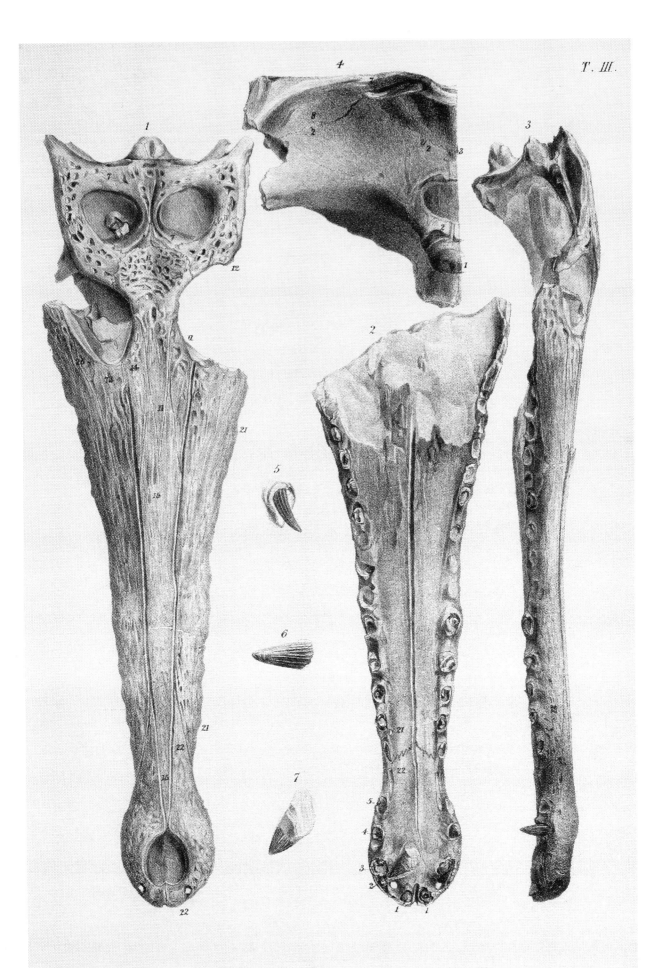

J. Erxleben

TAB. IV.

Vertebræ of the *Crocodilus toliapicus,* nat. size.

Fig.

1. Fore part of a mutilated seventh cervical vertebra.
2. Under part of ditto.
3. Side view of a mutilated eighth cervical vertebra.
4. Front view of the same vertebra.
5. Under view of the centrum of a lumbar vertebra.
6. Under view of the centrum of the fourth or fifth dorsal vertebra.
7. Side view of the first caudal vertebra.
8. Side view of a middle caudal vertebra.
9. Under view of the same vertebra.

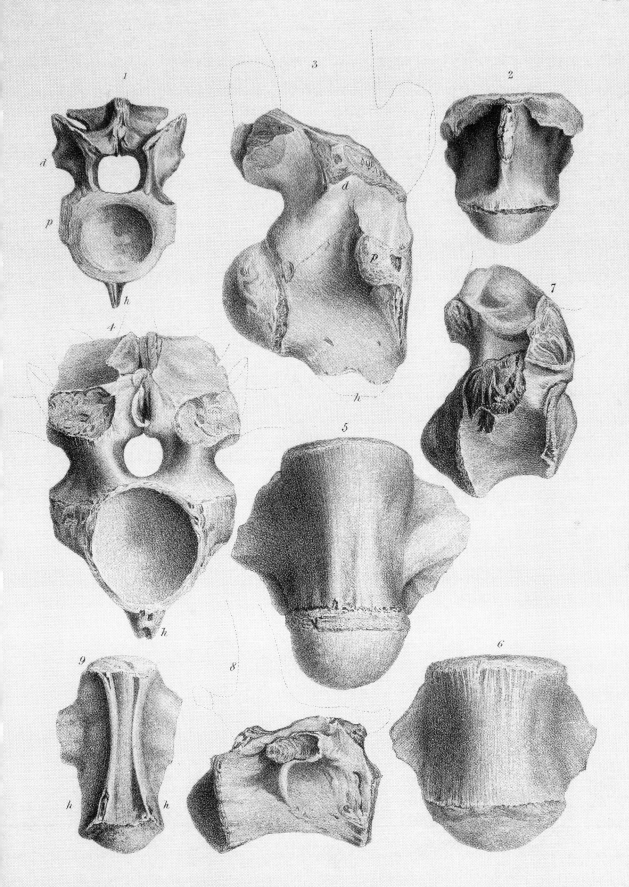

TAB. V.

Fig.

1. Side view of the fourth cervical vertebra (mutilated) of the *Crocodilus toliapicus*.
2. Under view of the same, showing the shape of the *hypapophysis h.*
3. Side view of the sixth cervical vertebra (mutilated) of ditto.
4. Side view of the first dorsal vertebra (mutilated) of the *Crocodilus champsoïdes*.
5. Front view of the anterior sacral vertebra (mutilated) of the *Crocodilus toliapicus*.
6. Side view of the same vertebra.
7. Side view of the third cervical vertebra (mutilated) of the *Crocodilus champsoïdes*.
8. Under view of the same vertebra.
9. Side view of a posterior dorsal vertebra (mutilated) of·the *Crocodilus champsoïdes*.
10. Side view of the centrum of the first caudal vertebra of ditto.
11. An anterior tooth of the *Crocodilus champsoïdes*.
12. A posterior tooth of the *Crocodilus toliapicus*.

All the figures are of the natural size.

TAB. VI.

CROCODILIA—*Crocodilus Hastingsiæ*, half the nat. size.

Fig.
1. Upper view of the skull.

2. Back view of ditto.

3. Ectopterygoids (25) and pterygoids (24), with the posterior aperture of the nostril.

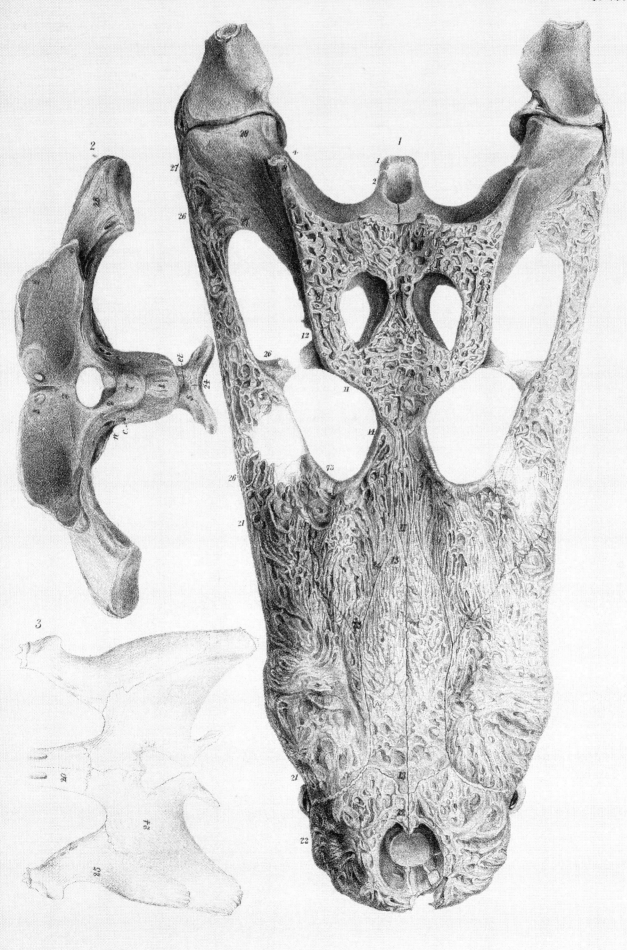

TAB. VII.

CROCODILIA—*Crocodilus Hastingsiæ,* half the nat. size.

Fig.
1. Side view of the skull.

2. Under view of the cranial and facial parts of ditto.

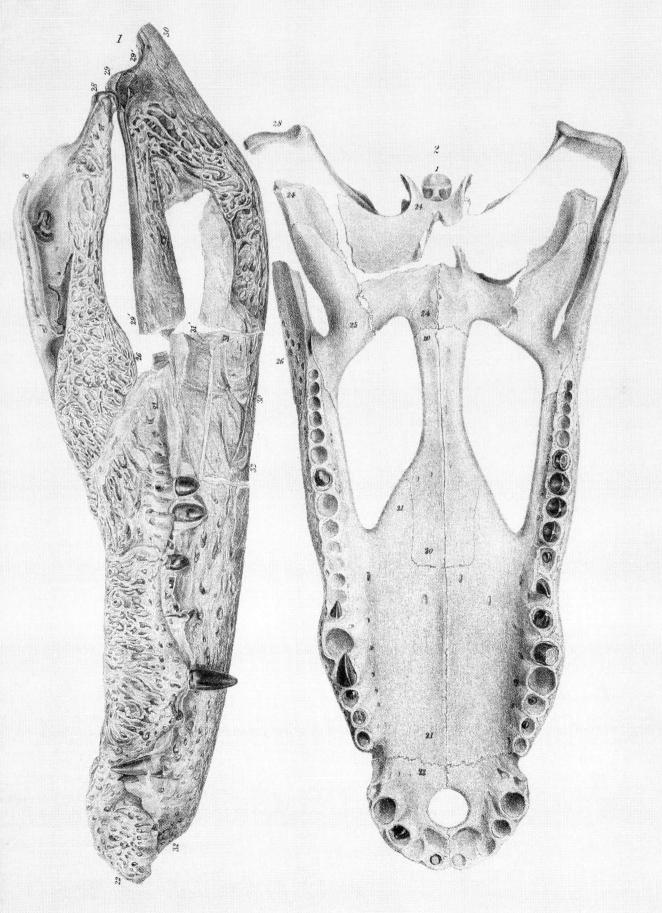

TAB. VIII.

Fig.

1. Under view of the fore part of the upper jaw of the *Crocodilus Hastingsiæ*, nat. size.

2. The same view of the *Alligator Hantoniensis*, nat. size.

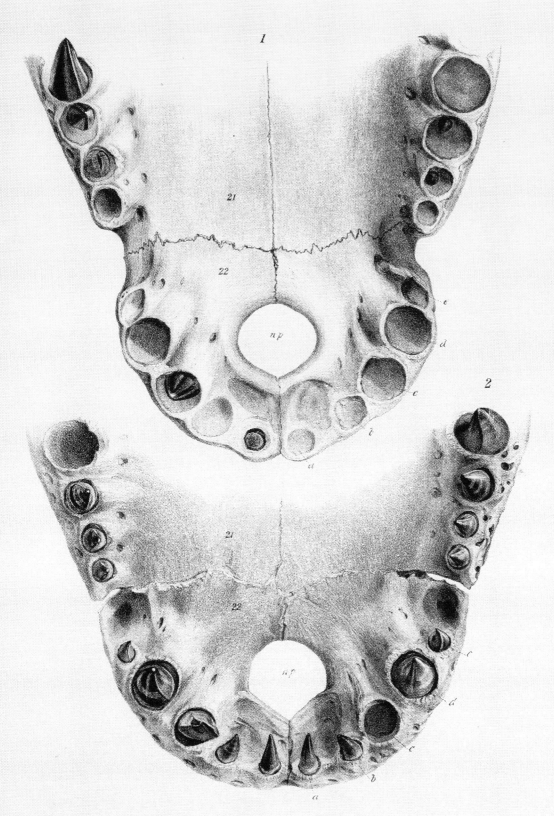

1. *Crocodilus Hastingsiæ.* 2. *Alligator Hantoniensis.*

TAB. IX.

Crocodilus Hastingsiæ, nat. size

Fig.

1. Side view of the fourth cervical vertebra.
2. Back view of the same vertebra.
3. Side view of a similar vertebra, *minus* the pleurapophyses.
4. Under view of the third cervical vertebra.
5. Back view of a lumbar vertebra.
6. Front view of the first sacral vertebra.
7. Side view of the first caudal vertebra.

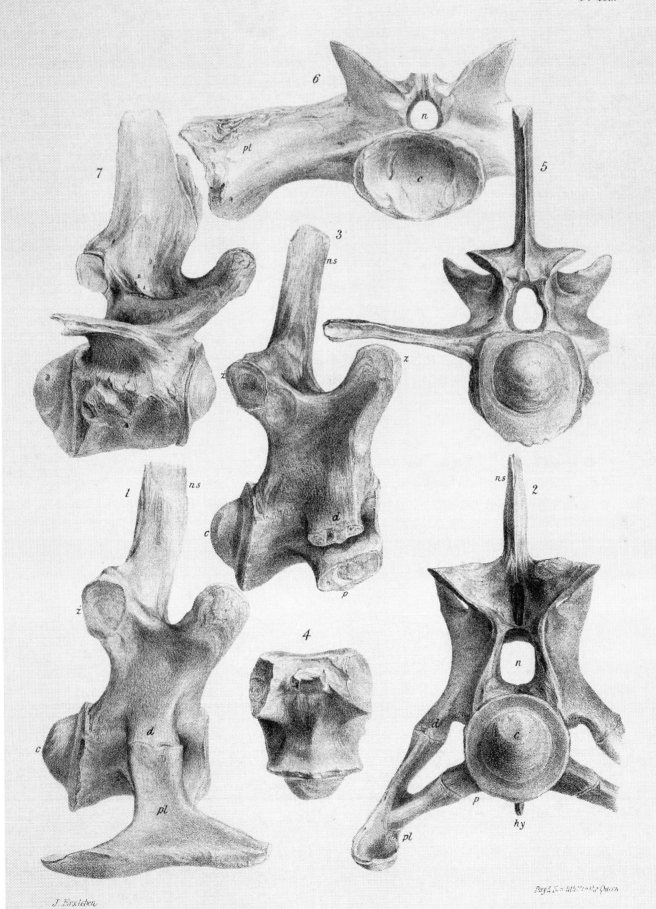

TAB. X.

Gavialis Dixoni, nat. size.

Fig.
1. A fragment of the symphysis of the lower jaw.
2. A fragment of one of the rami from the back part of the symphysis, showing the depressions *cc,* in the interspaces of the alveoli.
3. The fractured under part of a fragment of the same (*d*) jaw, exposing the hollow base of a young tooth.
4. A fragment of a ramus forming the long symphysis of the under jaw of a younger individual of the same species; upper surface.
5. Under side of the same fragment.
6. A portion of a tooth of the same species of Gavial.
7. A tooth with the germ of its successor, which has entered its base, of the same Gavial.
8. The centrum of a cervical vertebra of the same Gavial.
9. The femur of the same Gavial.
10. The crown of a tooth of a large Crocodilian from Bracklesham.

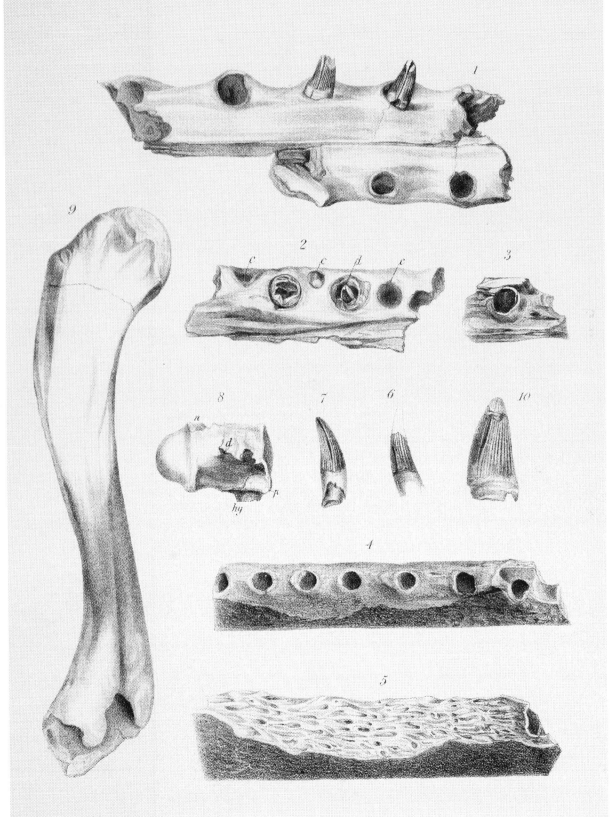

TAB. XI.

CROCODILIA.

Fig.

1. Side view of the skeleton of a Gavial.

1 *a*. Upper view of the skull of ditto.

2. Side view of the skeleton of a Teleosaur.

2 *a*. Upper view of the skull of ditto.

On the scale of one inch to a foot.

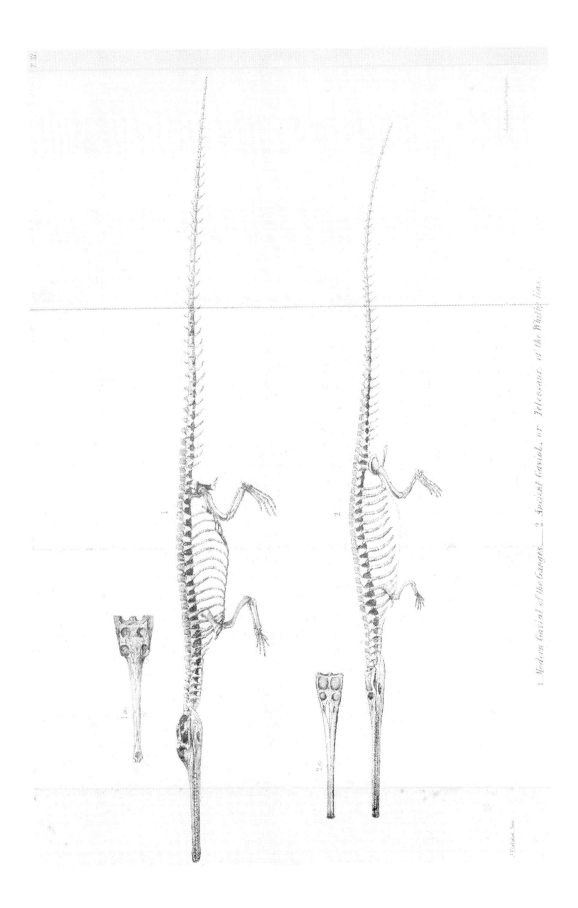

1 Modern Gavial of the Ganges.———2 Ancient Gaviol, or Teleosaur of the Whitby lias.

The material originally positioned here is too large for reproduction in this reissue. A PDF can be downloaded from the web address given on page iv of this book, by clicking on 'Resources Available'.

TAB. XII.

Fig.

1. Upper surface of the symphysial end of the right ramus of the lower jaw of *Crocodilus biporcatus*.

2. The same part of *Crocodilus Hastingsiæ*.

3. The same part of *Alligator niger*.

4. The hinder part of the right ramus of the lower jaw of *Crocodilus biporcatus*.

5. The same part of *Crocodilus Hastingsiæ*.

6. The same part of *Alligator niger*.

Figures 1, 2, 4, 5, two-thirds, 3 and 6 one half the nat. size.

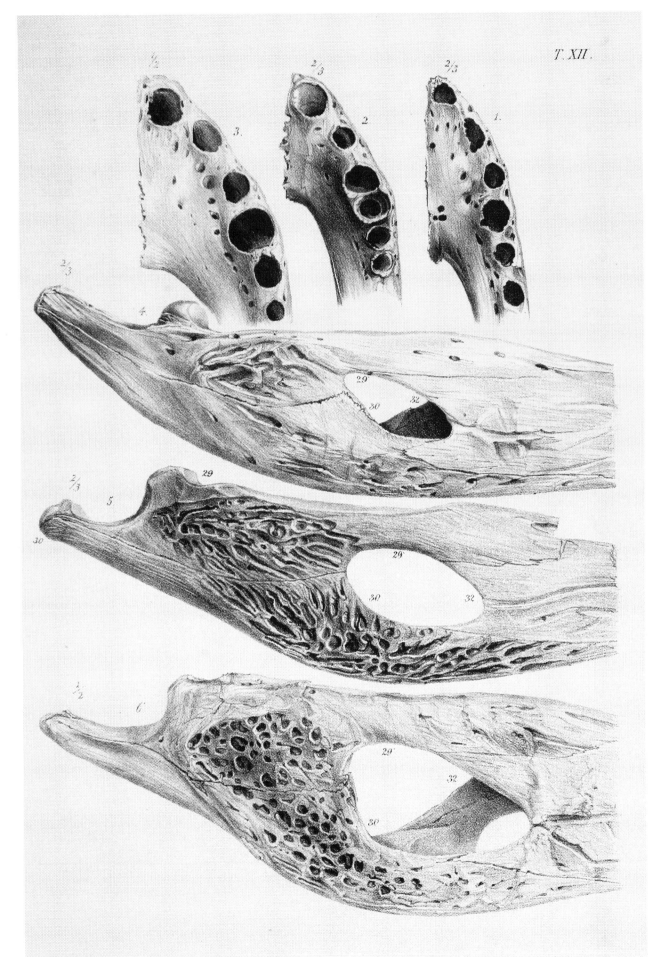

MONOGRAPH

ON

THE FOSSIL REPTILIA

OF THE

LONDON CLAY, AND OF THE BRACKLESHAM AND OTHER TERTIARY BEDS.

PART III.

PAGES 51—68; PLATES XII—XVI.

OPHIDIA (PALÆOPHIS, &c.).

BY

PROFESSOR OWEN, D.C.L., F.R.S., F.L.S., F.G.S., &c.

Issued in the Volume for the Year 1849.

LONDON:
PRINTED FOR THE PALÆONTOGRAPHICAL SOCIETY.
1850.

PART III.

Order *OPHIDIA.*

SERPENTS.

PRIOR to the publication of my Memoir on the *Palæophis* in the 'Geological Trans-actions,'* and my 'Report on British Fossil Reptiles,'† the sole notice of any fossil belonging to the order of Serpents was contained in the following passage from the Appendix to the concluding volume of the second edition of Baron Cuvier's great and comprehensive work, the 'Recherches sur les Ossemens Fossiles.' After alluding to the scarcity of the fossil remains of birds, the immortal author of that work proceeds to say: "The bones of Serpents are still rarer, if it be possible. I have seen no specimens of them, save the vertebræ from the osseous breccia of Cette, of which I have spoken in the article on those breccia, and a single one from the fresh-water deposits of the Isle of Sheppy."‡

We may perhaps gather the reason for the silence of Cuvier respecting the relations of that vertebra and of the fossil vertebræ of Serpents in general to each other, and to those of the existing species, from his brief notice of the Ophidian fossils from the breccia of Cette; where, after stating in general terms their resemblance in form and figure to the vertebræ of the common harmless snake (*Coluber natrix*), he proceeds to remark, "but it may well be conceived, that in a genus where the osteo-logy of the species has so much similitude, it is not in isolated vertebræ that one can discover specific characters."§ If, however, this discouraging conclusion of the great comparative anatomist should be countenanced by the results of a rigorous comparison of the vertebræ of the different species of *Coluber*, as that genus may be restricted by modern naturalists, it is by no means borne out by such comparison of the vertebræ of the species of the wider Linnean genus *Coluber*, and gives place to a very different estimate of the value of vertebral characters, when these are studied in species of the different Linnean genera of the '*Amphibia* Serpentes' in the 'Systema Naturæ.'

Baron Cuvier having, conformably with his convictions, deemed it unnecessary to give figures or to describe the vertebræ of Serpents, recent or fossil, in his 'Ossemens

---

* Vol. vi, 2d series (1839), p. 209, pl. xxii.

† Report of the British Association for the Advancement of Science, for 1841.

‡ "Les os de serpens sont encore plus rares, s'il est possible. Je n'en ai vu que des vertèbres des brèches osseuses de Cette, dont j'ai parlé à l'article de ces brèches, et une seule des terrains d'eau douce de l'île de Sheppy." (Tom. v, pt. ii, p. 526, 1824.)

§ "Mais on sent bien que, dans un genre ou l'ostéologie des espèces a tant de ressemblance, ce n'est pas dans les vertèbres isolées que l'on peut trouver les caractères spécifiques." (Op. cit., tom. iv, p. 180.)

Fossiles,' I am compelled to premise such observations on the anatomical construction of this part of the skeleton of those Reptiles as will render intelligible my description of the fossil ophidian vertebræ, and vindicate the grounds on which some of these are referred to distinct species, and others to genera of which we have no evidence of the actual existence in living Nature.

I have selected as the type of an ophidian vertebra that of a large, terrestrial, constricting Serpent (*Python Sebæ*), an African species, which makes the nearest approach in size to some of the fossil ophidian vertebræ from British tertiary strata. The vertebra figured in T. XIII, figs. 1-4, is from about the middle of the back of a specimen which was twenty feet in length. In the Pythons, as in other known *Ophidia*, all the autogenous elements, except the pleurapophyses (pl, figs. 2′, 3′), coalesce with one another in the vertebræ of the trunk; and the pleurapophyses (T. XIV, fig. 42, pl) also become anchylosed to the diapophyses (ib. d) in those of the tail. There is no trace of suture between the neural arch (T. XIII, figs. 1-4 n) and centrum (c). The outer substance of the vertebra is compact, with a smooth or polished surface. The vertebræ are procœlian, the cup (fig. 2 c) being deep, with its rim sharply defined and most produced at the sides; the cavity looking not directly forwards, but a little downwards, from the greater prominence of the upper over the lower border: the well-turned prominent ball (fig. 3 c) terminates the back part of the centrum rather more obliquely, its aspect being backwards and upwards. The hypapophysis (h) is developed in different degrees from different vertebræ, but throughout the greater part of the trunk presents the form and proportions shown in figs. 1, 4 h. A vascular canal perforates the under surface of the centrum (fig. 4), and there are sometimes two or even three smaller foramina. A large, vertically oblong, but short diapophysis (d) extends from the fore part of the side of the centrum obliquely upwards and backwards. It is covered by the articular surface for the rib, which is convex lengthwise, and convex vertically at its upper half, but slightly concave at its lower half. The base of the neural arch swells outward from its confluence with the centrum, and developes from each angle a transversely elongated zygapophysis; that from the anterior angle (z) looking upwards, that ($z′$) from the posterior angle downwards, both surfaces being flat, and almost horizontal. A thick rounded ridge connects the anterior with the posterior zygapophysis on each side, extending along the base of the neural arch. The neural canal (fig. 2, n) is narrow, with a subtriedral area, and with a narrow longitudinal ridge on each side. The neural spine (ns) is of moderate height, which scarcely equals its antero-posterior extent; it is compressed and truncate. A wedge-shaped process—the 'zygosphene'* (zs, fig. 2)—is developed from the fore part of the base of the spine; the lower apex of the wedge being, as it were, cut off, and its sloping sides presenting two smooth, flat, articular surfaces. This wedge is received into a cavity—the 'zygantrum'†

---

* Ζῠγόν, a yoke, σφήν, a wedge.   † Ζῠγόν, and ἄντρον, a cavity.

(fig. 3, *za*)—excavated in the posterior expansion of the neural arch, and having two smooth articular surfaces to which the zygosphenal surfaces are adapted.

Thus the vertebræ of Serpents articulate with each other by eight joints in addition to those of the cup and ball on the centrum; and interlock by parts reciprocally receiving and entering one another, like the joints called tenon-and-mortise in carpentry.

This is the most conspicuous, but is not the peculiar characteristic of an ophidian vertebra; the zygosphene (*zs*) and zygantrum (*za*) being developed in certain Lacertians, e. g. the genus *Iguana* (T. XIII, figs. 34, 35), but here the articular diapophysis (fig. 33, *d*) is much smaller, and forms a simple, convex, sessile tubercle; the hypapophysis is wanting: the zygosphene (fig. 34, *zs*) is deeply notched anteriorly, and the zygantra (fig. 35, *za, za*) are shallow, and separated from each other behind: a thick rounded eminence extends backwards from the diapophysis to the ball on the back part of the centrum (fig. 36); and that ball is a transverse ellipse (fig. 35), not hemispheroid, as in the *Ophidia*.

With regard to the specific distinctions which may be deduced from the characters of the vertebræ of Serpents, it is requisite first to determine the extent to which those characters vary in the vertebral column of the same species.

The atlas and axis are modified in the same degree as in the *Crocodilia*, with the addition of the entire suppression of their pleurapophyses. The atlas (T. XIV, figs. 38, 39) has two hypapophyses, one behind the other, as we shall find to be the case in other vertebræ of one of the great fossil Serpents. The normal hypapophysis (*h*, fig. 39), answering to that marked *ca, ex* in the woodcut, fig. 8, p. 85, is autogenous and wedge-shaped, as usual in the Reptilia; and is articulated on each side to a small portion of the neurapophysis (*n*); it also presents a concave articular surface anteriorly (fig. 38, *h*) for the lower part of the basioccipital tubercle, and a similar surface behind for the detached central part of the body of the atlas (fig. 40, *ca*), which is here confluent with that of the axis (fig. 40, *cx*), forming the so-called odontoid process of that vertebra; its Ophidian peculiarity being the development of an exogenous hypapophysis (*h'*) from its under and back part, like the posterior hypapophysis of the succeeding vertebræ.

The base of each neurapophysis of the atlas (fig. 38, *n*) has an antero-internal articular surface for the exoccipital tubercle, and a postero-internal surface for the upper and lateral parts of the odontoid (*ca*), besides the small median inferior facet for the detached hypapophysis (*h*): they thus rest on both the separated parts of their proper centrum. The neurapophyses expand and arch over the neural canal, but meet without coalescing. There is no neural spine. Each neurapophysis developes from its upper and hinder border a short zygapophysis (*z*): and from its side a still shorter diapophysis (*d*).

The axis or second vertebra of the trunk with the partially coalesced body of the atlas or 'odontoid,' is represented at fig. 40, T. XIV.

The odontoid presents a convex tubercle anteriorly, which fills up the articular cavity in the atlas for the occipital tubercle : below this is the surface for the detached hypapophysial part of the atlas (fig. 39, *h*) and above and behind it are the two surfaces for the atlantal neurapophyses ; the whole posterior surface of the odontoid is anchylosed to the proper centrum of the axis.

The neural arch of the axis developes a short ribless diapophysis from each side of its base ; a short zygapophysis (*z*), from each side of its anterior border ; a thick sub-bifid zygapophysis (*z'*) from each side of the posterior border, and a moderately long retroverted spine (*ns*) from its upper part. The centrum terminates in a ball behind, and below this sends downwards and backwards a long hypapophysis (*hx*).

In the skeleton of an African Constrictor (*Python regius*, Dum.), which measured 15 feet 6 inches in length, there are 348 vertebræ, of which the 279 following the atlas and axis support simple moveable ribs ; of these vertebræ about 70 anterior ones have long hypapophyses, as in fig. 4, *h*, T. XIV, which in the rest subside to the obtuse ridge and tubercle, as in fig. 1, *h*, T. XIII : the caudal vertebræ have not the ribs moveably articulated ; they are 67 in number ; of these vertebræ 56 have bifurcate hypapophyses as in fig. 42, *h*, T. XIV ; the six anterior caudals have bifurcate ribs (ib., fig. 41, *pl*), in the rest they are simple (ib., fig. 42, *pl*), and lengthen out the diapophyses (ib. *d*,) to which they are anchylosed.

The ribs or the trunk-vertebræ, like those of the tail, are 'pleurapophyses' or 'vertebral ribs ;' there are no 'hæmapophyses' or sternal ribs ; the exogenous hypapophyses (*h*, fig. 42) take the place of the hæmapophyses in the tail. The pleurapophyses of the trunk (T. XIII, figs. 2', 3'), are long, slender, curved, subcompressed, expanded at the proximal end, which presents an articular surface chiefly concave, and adapted to the diapophysial tubercle (*d*, figs. 2, 3,) above described ; there is a rough depression on the fore part of the expansion for the insertion of a ligament, and a tuberosity projects from the upper and back part ; the distal end of the rib is truncate, with a terminal pit ; a medullary cavity extends through a great part of the length of the rib, as shown in fig. 3''.

There is a small cavity in the substance of each neurapophysis, which communicates by a smaller foramen with the zygantrum. A vascular cavity in the centrum communicates with the neural canal.

In the skeleton of a Tiger-boa (*Python tigris*), in the Museum of the Royal College of Surgeons, measuring eleven feet in length, and having 291 vertebræ, the 253 following the axis support simple moveable pleurapophyses, articulated to concavo-convex sessile diapophyses, and constitute the dorsal, abdominal, or trunk-vertebræ ; 70 of the anterior of these vertebræ have long hypapophyses, as in fig. 4, *h*, T. XIV, they then begin to shorten, and subside to the ridge and tubercle, as shown in fig. 1 and 4, T. XIII, in the rest of the trunk-vertebræ. The first caudal vertebra has free pleurapophyses, but they are short and bifurcate, the upper prong being the shortest ; in the second caudal the left bifurcate rib is free, but the right is anchylosed to the diapophyses ; the prongs are of equal length in this and the two following vertebræ.

In the fifth caudal the outer prong is again shorter, and in the sixth it is a mere tubercle; at this part of the tail the hypapophyses begin to lengthen, bifurcate, and progressively increase in length to the sixteenth caudal, and thence gradually diminish and subside; yet the general configuration of the neural arch, the contour and degree of production of its posterior border, and the shape of the zygosphene, remain almost unaltered throughout.

In a true *Boa constrictor*, with 305 vertebræ, 71 at the anterior part of the trunk have long hypapophyses; and of the 60 caudal vertebræ, 44 have bifid hypapophyses. The first caudal is characterised by the sudden shortening of its ribs, and by a short process from the middle of their outer surface: this process is longer and nearer the head in the next rib; and in the third caudal vertebra the rib seems to bifurcate from its proximal end, which has become anchylosed to the diapophysis. Beyond the eighth caudal the outer costal prong or process disappears, and the anchylosed rib represents a long deflected diapophysis to within three or four vertebræ from the end of the tail. The last imperforate obtuse bone of the tail is obviously a coalescence of three vertebræ.

In the Rattlesnake (*Crotalus*, T. XIII, figs. 9-12) the hypapophyses (h) continue to be developed singly, and of equal length with the neural spines (ns), throughout the trunk; and any single vertebra might be distinguished from an anterior trunk-vertebra of a *Boa* or *Python* by the following characters: the diapophysis (d) developes a small, circumscribed, articular tubercle from its upper convexity, and a short process (d') from its under part, extending downwards and forwards below the level of the centrum (c); the anterior zygapophysis (z) seems to be supported by a similar process (d'') from the upper end of the diapophysis, the point of which projects a little beyond the end of the zygapophysis (fig. 10); the zygapophyses are less produced outwards than in the Python; the zygantra (za, fig. 11) are more distinct excavations.

In the Cobra di Capello (*Naja*, T. XIII, figs. 13-16), the diapophysis presents the same well-marked tubercle (d) upon its upper part, but its lower end (d') is much less produced than in the Rattlesnake; the process of bone (d'') underpropping the zygapophysis projects proportionally further beyond the articular surface (z): the neural spine (ns) is much lower, and beyond the anterior third of the trunk the hypapophysis (h) subsides into a ridge, with its point produced backwards beneath the articular ball of the centrum; the zygantra (za, fig. 15) are distinct cavities.

In the *Coluber elaphus* (T. XIII, figs. 17-20) the trunk-vertebræ are distinguished by the great extent to which the part of the diapophysis (d'', fig. 18) which underprops the zygapophysis (z) is produced beyond the articular surface, the lower end of the diapophysis (d') is less produced; the hypapophysis, beyond the anterior fourth part of the vertebral column, is reduced to a straight ridge, (fig. 20, h), extending along the middle of the under surface of the centrum, and not produced posteriorly: a groove separates the ridge on each side from the diapophysis and the posterior ball of the centrum. Both the cup and ball and the articular part of the diapophysis are relatively smaller than in the *Naja;* the neural spine (*ns*, fig. 17) is lower in proportion to its antero-posterior extent. The pleurapophysis (*pl*) is shown articulated to the tubercle in figs. 19 and 20.

The vertebræ of the common harmless Snake, *Coluber natrix,* differ only in size from those of the larger continental species above described.

In an African *Eryx* (T. XIII, figs. 21-24) the diapophysis (*d*) does not extend beyond the articular surface of the anterior zygapophysis (*z*), but is exclusively devoted to forming a low, subconvex, articular tubercle, which has a longitudinal depression anteriorly; the posterior margin of the neurapophysis (fig. 21, *n*) forms an angle above the zygantrum, which angle, though slight, is more marked than in any of the foregoing Ophidians; the hinder end of the hypapophysial ridge (*h*) is slightly produced; the zygapophyses (*z, z'*, fig. 24) are less extended outwards than in the Pythons.

In a Sea-snake (*Hydrophis bicolor*, T. XIII, figs. 25-28) I find the height of the neural spine (fig. 25, *ns*) greater in proportion to its antero-posterior extent than in any of the foregoing Ophidians. The diapophysis (*d*) sends a point (*d''*) outwards a little beyond the articular surface of the anterior zygapophysis (*z*); a very small hypapophysis (*h*) projects below the articular ball of the centrum, and a low ridge is continued forwards from it (fig. 28); the posterior border of the neurapophysis (fig. 25, *n*) forms no angle, but is moderately convex, as in all the foregoing Ophidians, excepting the *Eryx*.

With this indication of the kind and extent of the vertebral characters of the different species of Serpent which I have been able to study in reference to the fossils to be described, I proceed to the comparisons by which the following extinct genera and species have been established.

Genus—Palæophis.

Palæophis Typhæus, *Owen.* Tab. XIII, figs. 5-8, Tab. XIV, figs. 1-3, 7-9, 16, 17, 26, 27, 28. (Fig. 6, 10-12)?

Amongst the numerous vertebræ of this species of Serpent which have come under my examination, a few, of small size, have shown the hypapophysis long and compressed, as in the specimen in Mr. Bowerbank's collection, figured in T. XIV, figs. 1-3, *h*, indicating that the vertebræ at the anterior part of the trunk had that character, as in the large existing Serpents; whilst all the larger vertebræ, with the hypapophysis perfect, manifest shorter proportions of that process, as in the typical example, apparently from the middle of the abdomen (T. XIII, figs. 5-8, *h*); whence I infer that the *Palæophis* resembled the *Python, Boa, Coluber,* and *Hydrus*, in having different proportions of the hypapophysis at different parts of the vertebral column. Had every fossil vertebra shown a long hypapophysis like that in T. XIV, fig. 1, we might have suspected that the species had been of the venomous family, like the Rattlesnake.

The veritable Ophidian nature of the fossils in question is demonstrated, not only by the superadded zygosphenal (*zs*, fig. 6) and zygantral (*za*, fig. 7) articulations, but by the solidity of the zygosphene, by the size and form of the centrum, by those of its articular cup (*c*, fig. 6) and ball (*c*, fig. 7), and of its hypapophysis (*h*); and also by the size and prominence of the diapophysis (*d*). The largest vertebræ (e. g. T. XIII, figs. 5-8, and T. XIV, figs. 16, 26, 27, 28) probably from about the middle of the body, as compared with the vertebræ from the same part of the skeleton of a *Python Sebæ*, twenty feet in length, are longer in proportion to their breadth, and the cup and ball of the centrum are larger; the hypapophysis (*h*) is more produced, and there is a second smaller hypapophysis close to the anterior part of the under surface of the centrum, which in most of the large vertebræ is connected by a ridge with the hinder and normal hypapophysis; but in a few vertebræ is not so connected. The articular cup and ball are less obliquely placed upon the extremities of the centrum, being nearly vertical (compare fig. 5 and fig. 1, *c'*). The rim of the cup is sharply defined, and is more produced from between the bases of the diapophyses; a deeper and narrower chink intervening than in the *Python*. The transverse diameter of the cup (*c*, fig. 6) is greater than that of the zygosphene (ib., *zs*)—a proportion which I have not found in the vertebræ of any existing genus of Serpent, in which the base of the zygosphene always equals at least the parallel diameter of the articular cup. The articular part of the diapophysis is more produced outwards and less extended vertically in *Palæophis* than in *Python*, and it is uniformly convex; a ridge is continued from its upper end obliquely forwards to, but not beyond, the apex of the anterior zygapophysis (*z*), forming the angle between the lateral and anterior surfaces, whilst the horizontal articular facet forms the third surface of that three-sided conical process. In the Python the non-articular part of the same zygapophysis is convex, and the process is much more extended outwardly; the proportions of the zygapophysis in the *Palæophis* more resemble those in the *Coluber* and *Hydrus*, but differ from these, as also from *Naja* and *Crotalus*, in the non-extension of the diapophysial point beyond the articular surface.

A ridge or horizontal rising of the bone extends from the anterior to the posterior zygapophysis, but is more or less blunted or subsides midway, and is by no means so produced outwards as in Python; in this respect more resembling that in *Coluber* and *Hydrus*. Below the middle of this ridge, on a level with the upper surface of the centrum, there is a short, nearly parallel rising in *Palæophis* (fig. 5). The zygosphene (fig. 6, *zs*) is slightly excavated anteriorly, and shows no trace of the tubercle which characterises the middle of that surface in the Python (fig. 2); it is also broader in proportion to its height. But perhaps the most characteristic feature of the vertebra of the *Palæophis* is the peculiar production of the posterior border of the neurapophysis into an angle (*n*, fig. 5) directed upwards, outwards, and backwards, and this is common to all the species; there is no trace of this process in the *Hydrus* (fig. 25), and the nearest approach to it which I have hitherto met with among existing

8

Serpents, is that low, tuberous angle at the corresponding part of the vertebra of the *Eryx* (fig. 21). The posterior zygapophysis resembles, of course, the anterior one in its much less extent, especially transversely, as compared with that in the Python, and the posterior border of the neurapophysis (fig. 5, *z'*, *n*) rises from its apex vertically, or a little inclined outwards and backwards, giving a squarish form to the surface of the neural arch in which the zygantra (*za*, fig. 7) are excavated; these cavities, in proportion to the articular ball beneath, are smaller and less deep than in the Python, or any other existing genus of Serpent. The sloping sides of the neural arch above the zygapophysial ridge are more concave than in Python, and so resemble those parts in *Coluber* and *Hydrus*. The latter genus (fig. 25) and *Crotalus* (fig. 9) most resemble *Palæophis* in the proportions of the neural spine (*ns*); this part, however, in *Palæophis* differs from that of *Hydrus* in having its base coextensive with the supporting arch, springing up from the fore part of the zygosphene, whilst this part entirely projects forwards, clear of the base of the spine in *Hydrus*, as in *Python*, *Coluber*, and *Naja*; but in *Crotalus* the base of the spine has the same antero-posterior extent as in *Palæophis*, and it comes very near to the fore part of the zygosphene in *Eryx*. The neural spine has been more or less fractured in every specimen of the brittle crumbling vertebræ of the *Palæophis Typhæus* from the Bracklesham Clay; only one specimen, which I carefully worked out in relief from a mass of matrix, after imparting some of its original tenacity to the substance of the bone, affords a true idea of the peculiar character of these Ophidian vertebræ, which is afforded by the great height of the neural spine (see T. XIV, fig. 27, *ns*); but even here, although the fore part of the spine equals in vertical extent that of the rest of the vertebra beneath it, I am not sure that its entire extent is preserved, the part having been obliquely broken away behind this point before the specimen came into my hands. Some vertebræ of another species of *Palæophis* from Sheppy, show this elevated spine to be a generic characteristic of the fossil vertebræ.

	Inches.	Lines.
The entire vertical extent of the vertebra, fig. 27, from the end of the hypapophysis to the summit of the neural spine is .	2	5
The length of the same vertebra is	1	2
The length of a larger vertebra of the same species . .	1	4
The length of the smallest free vertebra	0	5

I have specified this last vertebra as being 'free,' because in Mr. Dixon's collection, by far the richest in the remains of the great *Palæophis* of Bracklesham, there are two smaller vertebræ anchylosed together by both their bodies and neural arches (T. XIV, figs. 32, 33, 34), which, therefore, are not 'atlas and axis,' but from their compressed form I should judge rather to have come from the opposite end of the vertebral column: they have not formed, however, the very extremity of the tail, like the terminal anchylosed vertebræ in the *Boa constrictor*, or those supporting the rattle in the *Crotalus;* for the ball of the centrum, the posterior zygapophyses, and the zygantral articulations, are present on the back part of the second of these anchylosed vertebræ. But before further pursuing the description of this remarkable specimen, I shall premise a brief notice of the vertebræ which have presented other modifications.

In the series of Palæophidian vertebræ from Bracklesham, which I have had the opportunity of comparing, a few, as has been already remarked, appear to have come from the fore part of the body by the length of the hypapophysis, as compared with the size of the vertebræ, which is small; a character that adds to the likelihood of their having come from that extremity of the series. Fig. 1, T. XIV, shows one of these vertebræ in which the hypapophysis (*h*) is entire; it is shorter and much more compressed than that process is in the anterior trunk-vertebræ of the Python, fig. 4, and its base extends forwards, as a sharp ridge, to between the diapophyses (fig. 3), where like them, it has been mutilated by fracture. The zygapophyses are small, and there is no ridge continued from the anterior to the posterior one; the neurapophysis presents the characteristic angular production (fig. 1, *n*), and the neural spine (ib., *ns*) is coextensive with the supporting arch.

The second form of vertebra is characterised by a single and moderately-developed hypapophysis, the base of which is confined to the hinder half of the under surface of the centrum, leaving the fore part of that surface concave where it expands between the bases of the diapophyses. I have received twelve such vertebræ from Bracklesham, varying in size between the two extremes given in figs. 5, 6, and 10, 11, 12, T. XIV. The hypapophysis (*h*) which is best preserved in the vertebra fig. 10, is shorter but thicker than in fig. 1; the articular cup and ball are relatively smaller; the zygosphene (*zs*) is larger, and its surfaces larger and more vertical (fig. 5); the neural spine has a less antero-posterior extent. These may be vertebræ from the hinder part of the abdomen, near the beginning of the tail. Some of them have a minute ridge at the middle of the anterior inferior concavity (fig. 9).

A third modification of vertebra shows the same limited extent of the base of the posterior hypapophysis, but a second shorter hypapophysis is constantly developed from the middle of the space between the bases of the diapophyses. I have examined twenty of such vertebræ ranging in size between the extremes given in figs. 14, 15, and 17, T. XIV. As compared with fig. 5, the articular cup of fig. 14 is larger, the zygosphene less, and of a different shape, concave anteriorly and not straight above, but forming an obtuse angle there. A ridge is continued from the posterior to the anterior zygapophysis (fig. 13).

This ridge is more strongly developed in the larger vertebræ with the same modification of the under surface (figs. 20, 21). The articular ball of the diapophysis would seem not to have descended so low down as in the typical vertebræ referred to *Pal. Typhæus.* The neural spine does not extend to the fore part of the zygosphene; there is a short but well-defined space above zygosphene in front of the spine.

Figs. 18, 19, 20, 21, give views of two of the best-preserved vertebræ of the present form, which I have attributed to a distinct species under the name of *Palæophis porcatus*.

The fourth modification of the Palæophidian vertebræ from Bracklesham is the most common, and is characterised by the coextension of the base of the hypapophysis with the under surface of the centrum, or by the whole of the middle of that under surface forming a ridge : both ends of the ridge being produced, the posterior one the most, and forming the normal hypapophysis. These vertebræ are usually of large size; I have examined upwards of thirty, ranging between the extremes given in figs. 25 and 26, T. XIV; and it is from this series that I have selected the type vertebra of the genus *Palæophis*, T. XIII, figs. 5-8.

The ridge between the anterior and posterior zygapophyses in these vertebræ is absent (T. XIII, fig. 5) or interrupted (T. XIV, figs. 27, 28). There is no well-defined space above the zygosphene anterior to the base of the neural spine. These vertebræ I regard as typical of the species *Palæophis Typhæus*: they are rather longer in proportion to their breadth than those of the *Palæophis porcatus*.

To this category belongs the vertebra with the unusually well-preserved neural spine (fig. 27), and likewise the two vertebræ which are preserved in their natural connexion, showing the reciprocal interlocking of their complex articular processes (fig. 28).

The fifth form of the vertebræ from Bracklesham is characterised by the compression of the centrum and the convergence of its almost flattened sides to the ridge on the inferior surface, from which a single hypapophysis is developed. I have examined not more than four such vertebræ, including the two which are anchylosed together, those (figs. 32-34, T. XIV) being the smallest in size, and the vertebra (figs. 29-31, T. XIV) the largest. The ridge between the anterior and posterior zygapophyses is suppressed; the neural arch gently swells out as it descends from the base of the neural spine, and from between the zygapophyses it bends in to coalesce with the converging sides of the centrum. This vertebra has not that character of a caudal vertebra, which is manifested in the *Python* and most modern *Ophidia* by the transverse pair of hypapophyses; it shows plainly the base of a single median hypapophysis from near the posterior surface of the centrum (fig. 34). The diapophyses of fig. 29 are broken away, together with the anterior concave end of the centrum; had they been entire, we might have derived from them evidence of the more constant character of the caudal vertebræ of Serpents, which is derived from the coalescence of a short and straight pleurapophysis with the diapophysis, lengthening out that transverse process, as in fig. 42. The zygosphene and zygantra are developed, as, indeed, they continue to be to near the end of the tail in modern Serpents; and the produced angle of the posterior border of the neurapophysis is as characteristic of the small compressed vertebræ of the *Palæophis* (fig. 29) as of the larger specimens.

The two anchylosed vertebræ belonging to the compressed series have been already alluded to. The base of the neural spine is limited to the posterior half of the neural arch in both (fig. 33). The hindmost of the two vertebræ is the longest, measuring five lines, the length of the two being nine lines. In each, the sides of the centrum are nearly plane, and converge at an acute angle to a ridge, which forms the under surface; a very small hypapophysis was continued from the back end of the ridge. This process is broken away from each vertebra, as are also the diapophyses, which are indicated by their rough fractured base; they are situated near the lower part of the side of the centrum, like the long diapophyses of the posterior caudal vertebræ of the Pythons; had they been preserved, their proportions would have determined whether the anchylosed vertebræ were caudal or not.

In the skeleton of a Tiger-boa (*Python tigris*) in the museum of the Royal College of Surgeons, anchylosis of the 148th to the 149th vertebra has taken place; and the 166th and 167th vertebræ have been more completely and abnormally fused together, so as to appear like a single vertebra on the left side, and a double one on the right side, where there are two diapophyses and two ribs. The compressed form, however, and diminutive size of the two anchylosed vertebræ of *Palæophis*, strongly indicate them to be from near the end of the tail, in which case it must be concluded that that part was compressed, as in the smaller modern *Hydrophides*, and that the present extinct Ophidian was a Sea-serpent of at least twenty feet in length.

All the vertebræ with the characters specified in the description of the large specimens from the trunk, and referable to the *Palæophis Typhæus*, have been obtained from the Eocene clay at Bracklesham, Sussex: they form part of the collections of the late Frederic Dixon, Esq., F.G.S., of Worthing; of James S. Bowerbank, Esq., F.R.S.; and of George Augustus Combe, Esq., of Preston, near Arundel, to whom I have been indebted for some beautiful examples, including the two vertebræ in natural conjunction (T. XIV, fig. 28), and the vertebra with the best preserved neural spine (ib., fig. 27.)

PALÆOPHIS PORCATUS, *Owen.* Tab. XIV, figs. 13-15, 18, 20, 21.

On comparing together eighteen Palæophidian vertebræ of different sizes from Bracklesham, the smallest of the dimensions represented in figs. 14, 15, and thence gradually increasing to the size of the specimen fig. 20, I find the following differences: in fig. 14, e. g. the articular cup and ball at the ends of the centrum are larger in proportion to the length of the centrum, as compared with the next-sized vertebra, fig. 5: the under surface of fig. 15 is convex transversely between the diapophyses and sends down a short median ridge; in fig. 6 it is concave at the same part, and without the median ridge; but both vertebræ have the median process or 'hypapophysis' at the back part of the under surface. In fig. 14 the fore part of the zygosphene is concave, in fig. 5 it is flat; in fig. 5 the upper border is straight, in fig. 14 it forms an open angle; the space between the zygosphene and zygapophysis is greater in fig. 5 than in fig. 14.

Twelve vertebræ of progressively increasing size repeat the characters of the vertebræ (fig. 6); i. e. they have the fore part of the under surface between the diapophyses excavated, and have only one inferior spine, viz. the hypapophysis developed from the hind part of the under surface; they have also the zygosphenal articulations nearly vertical, and raised high above those of the zygapophyses (fig. 8). A vertebra (figs. 22, 23, 24) of the same size as the largest of these twelve differs from them, and repeats the general characters of the small vertebra (fig. 14): it has the anterior as well as the posterior hypapophysis; larger terminal cup-and-ball surfaces in proportion to its size; smaller intervals between the zygosphenal and zygapophysial articulations (fig. 24); less lofty posterior aliform extensions of the neural arch, and the base of the neural spine extending nearly to the fore part of that arch. These vertebræ, and especially the larger specimens (figs. 18, 20) have a strong external ridge extending from the anterior to the posterior zygapophyses on each side of the neural arch. On comparing one of these vertebræ with another of the ordinary character and of the same size, the following further differences presented themselves: in the ridged vertebræ, which are provisionally referred for the convenience of description and comparison to a distinct species, with the name of *Palæophis porcatus*, the articular ball is broader in proportion to its height (compare fig. 23 with fig. 27); the anterior zygapophyses are more produced outwards and less produced forwards, so that they do not extend beyond the border of the articular cup, so far as in the non-ridged vertebræ of *Palæophis Typhæus*; the fore part of the zygosphene in the ridged vertebræ is broader, and less excavated. The breadth of the base of the neurapophysis is greater in the ridged vertebræ than in the unridged ones, in proportion to its length. The articular surfaces of the zygapophyses are smaller in the ridged than in the unridged vertebræ.

Figs. 13, 18, 20, 22, T. XIV, show the ridged character of the sides of the neural arch in *Palæophis porcatus*, and fig. 19 shows the consequent superior breadth of the base of that arch in relation to the length of the vertebra as compared with fig. 8, T. XIII, a corresponding vertebra of the *Palæophis Typhæus*. Fig. 14 in the same Plate shows the striking difference in the proportions of the same part of the vertebra in the *Python tigris*.

Such are the observed differences which seemed worthy of mention in the series of Palæophidian vertebræ from the Eocene deposits at Bracklesham which I have had the opportunity of comparing. The nature of the differences may be interpreted in different ways: with regard to the small vertebræ, for example, those with a single spine from the posterior part of the under surface (figs. 1, 2, 3, T. XIII) may be small cervical vertebræ of the same species as that to which the large vertebræ with the two inferior spines belong; and the small vertebræ with two inferior spines (figs. 14, 15) may have belonged to a smaller and younger individual of the same species, and have come from a more posterior part of the vertebral column of such individual. The anterior vertebræ of both Pythons and Boas, for example, are distinguished by an inferior spine, the remaining vertebræ to the tail being merely ridged beneath: but I have not met with such modifications in the trunk-vertebræ of the same existing Serpent, as those that have been pointed out in the vertebræ (figs. 5, 6, and figs. 14, 15); and in no specimen of Python or Boa, have I found the vertebræ presenting such differences of character as those indicated in the larger fossil Palæophidian vertebræ which I have described as 'ridged' and 'not ridged.' Leaving therefore the question of the nature of the differences in the smaller vertebræ (figs. 1 and 14) open, and as possibly depending upon difference of age and position in the series, I believe the characters of the ridged vertebra to be those of a distinct species of *Palæophis*.

Masses of mutilated vertebræ and ribs, irregularly cemented together by their matrix, are occasionally though rarely discovered in the Eocene clay at Bracklesham. The specimens of such aggregates which I have as yet seen have not exhibited any vertebræ sufficiently complete to yield more than the means of determining the generic relations. That of which a small portion is figured in T. XVI, fig. 4, is the most instructive, since it shows the form and structure of the ribs. The proximal half of the pleurapophysis (*pl*) equals in size the corresponding part in the *Python regius* of twenty feet; it shows the same fine cancellous structure of the articular end, and a similar medullary cavity, with thin compact walls, forming the body of the vertebra. The more slender distal portion of another rib is preserved, with the medullary cavity exposed at its fractured parts.

PALÆOPHIS TOLIAPICUS, *Owen.* Tab. XV and XVI.

> Transactions of the Geological Society of London, vol. vi, part ii, p. 209.
>
> Report on British Fossil Reptiles, in the Report of the British Association, 1841, p. 180.

The fossil Ophidian vertebræ which have been discovered in the London clay at Sheppy are, for the most part, smaller than those from Bracklesham; their common dimensions equalling those of a *Boa constrictor* of from ten to twelve feet in length. They all repeat, however, the generic modifications characteristic of *Palæophis;* the hinder margin of the neurapophyses (T. XV, fig. 5) is produced into a pointed or angular plate; the articular prominence for the rib (ib., fig. 3, *d*) is wholly convex; the zygapophyses are short, and no diapophysial point extends beyond the anterior ones; the height of the neural spine (T. XV, fig. 1, and T. XVI, fig. 2, *ns*) exceeds its antero-posterior extent. The veritable Ophidian character of the Reptile to which these fossil vertebræ belonged, is not only shown by their individual structure, but is well illustrated by the number of them in natural articulation which have occasionally been found cemented together in the petrified clay.

One of these Ophidiolites from the clay of Sheppy, in Mr. Bowerbank's collection, exhibits a portion of the vertebral column of the *Palæophis* suddenly bent upon itself, and indicating the usual lateral flexibility of the spine: in another specimen, including about thirty vertebræ, the vertebræ have been partially dislocated and are bent in a semicircle, Tab. XVI.

As compared with either of the species of *Palæophis* from Bracklesham, the vertebræ from Sheppy have the centrum proportionally longer and more slender, with a smaller terminal cup and ball. In vertebræ from Sheppy and Bracklesham in which those articulations were of equal size, the length of the neural arch at and including the zygapophyses, was two centimetres in the *Palæophis toliapicus*, and one centimetre, seven millimetres in the *Palæophis Typhæus*.

The hypapophysial ridge is more constant and better marked; it is produced at both extremities, and most so at the hinder one, but here in a less degree than in the *Palæophis Typhæus*, or *Pal. porcatus*, and the ridge is not interrupted between the two hypapophyses, as in most of the large vertebræ of the *Palæophis porcatus*. On the other hand, the rising of the bone continued from the anterior to the posterior zygapophysis does subside midway more completely than in the *Palæophis Typhæus;* and the ridge, which in that species extends to the apex of the produced posterior border of the neurapophysis along the outside of that aliform production, is less developed in the *Palæophis toliapicus:* the neural arch is less suddenly compressed above, or inclines more gradually to the base of the spine; this spine, also, although its base is extended to near the anterior border of the zygosphene, appears to be higher in proportion to its antero-posterior extent than in the *Palæophis Typhæus*. The diapophysis is less produced outwards and downwards than in the *Palæophis Typhæus* or *Palæophis porcatus*. In a group of thirty vertebræ of this species cemented together by the indurated clay from Sheppy, in the Hunterian Collection, and which, in the original MS. Catalogue of that part of John Hunter's Collection, were called 'vertebræ of a Crocodile,' T. XV, fig. 1, several of the long and slender subcylindrical ribs are also preserved, in the fractured parts of which the medullary cavity is shown. The articular surface at the proximal end presents the uniform concavity suited to the convexity of the diapophysis. I have seen no evidence of the process from the upper and back part of the proximal end of the rib which is present in the Python.

The finest and most strikingly Ophidian example of the great fossil snake of Sheppy has been obtained from that locality by Mr. Bowerbank since the publication of the Memoir in which his earlier specimens of the *Palæophis toliapicus* were determined and described. It consists of a series of thirty vertebræ, from about the middle of the abdomen, bent into an oval form upon their dorsal aspect, and measuring twenty inches in length (T. XVI, figs. 1, 2).

As the strong and complex articulations of these vertebræ in Serpents opposes any inflection of the column except from side to side, their unnatural bend in the fossil is attended with just the amount of mutual dislocation that was requisite to admit of it; but beyond this amount of dislocation, which chiefly affects the terminal ball and socket-joints, the vertebræ have been preserved in their natural juxtaposition and succession. The dead body of this eocene serpent has apparently sunk or been washed into the great stream or estuary, where it has been driven about to and fro, and variously contorted as it was swept along by the current; the portion here preserved has been by some external influences obstructed and bent upon itself in its present unnatural curve, as it finally sank in the sediment in that state of decomposition when the ligaments were ready to give way to the strain upon them; but the tough integuments, which have longer resisted dissolution, have served to retain the partially-dislocated vertebræ together until they became fixed in the matrix in the position in which they are now fossilized. We have in this condition very good evidence of that long and slender form of body which would admit of such an extent of inflection from external pressure in a direction contrary to that which the natural articulations of the vertebræ would allow; but since in Serpents those articulations are so strong, when fresh, as to offer considerable obstacles to any vertical inflection upwards or downwards, we may infer that the body of the *Palæophis,* of which the example in question formed a part, must have floated long enough to have undergone that degree of internal decomposition, which allowed it easily to yield to external pressure in any direction.

The characteristically long and comparatively slender spine is well preserved in the vertebræ at *ns,* fig. 2; and the equally characteristic angular production of the hinder border of the neural arch is shown in some other of the vertebræ. In many vertebræ the ribs are preserved just in that degree of juxtaposition in which they would remain after yielding to the pressure and movements of the overlying and accumulating sediment upon the integument of the body. Fig. 3 shows a portion of the coil, in which a few of the ribs offer to our view the concave articular surface (*pl*), which was articulated with the diapophysial tubercle (*d*): in fractured portions of the ribs their medullary cavity is shown. The hypapophysis which terminates the thick and low inferior ridge of the vertebræ of the present species offers that small degree of development characteristic of the middle and posterior part of the long abdominal region.

In T. XVI, fig. 1 shows this remarkable chain of vertebræ from the right side; fig. 2 the middle portion of the same chain from the left side; and fig. 3 the under surface of the vertebræ with the juxtaposed ends of the ribs.

In T. XV, fig. 1 shows a group of the vertebræ of *Palæophis toliapicus,* in some of which the long and slender spine (*ns*) characteristic of the genus is well preserved. In fig. 3 of the same plate, the position and form of the diapophysial tubercle (*d, d*) are shown. The character of the under surface of the vertebræ is shown in fig. 4, and the angular aliform production of the neural arch is shown in fig. 5.

One of these characteristic examples of the *Palæophis toliapicus* is preserved in the Hunterian Museum*, the others in that of James S. Bowerbank, Esq., F.R.S. In the Museum of Mr. Saull, F.G.S., a few vertebræ, and a fragment of the skull of probably the same species of *Palæophis,* likewise from Sheppy, are preserved.

---

* This was figured, by the permission of the President and Council of the College, in illustration of my original Memoir on the Genus *Palæophis* in the Geological Transactions.

On a general review of these numerous and rich accessions to our previously scanty evidence of extinct Serpents, I may sum up by stating that the generic character of *Palæophis* is chiefly manifested in the length of the neural spine, in the pointed aliform productions of the back part of the neurapophyses, in the uniform convexity of the diapophysial tubercles, and the minor transverse production of the zygapophyses.

The *Palæophis toliapicus* is distinguished by its longer vertebræ in proportion to their breadth, by its sessile diapophyses, and by the carinate character of the lower part of the centrum in the vertebræ of the abdomen.

The *Palæophis Typhæus* is distinguished by its shorter and broader vertebræ, by its pedunculate diapophysis, and by the anterior and posterior hypapophyses of the vertebræ of the abdomen ; its neural arch is narrower, and its sides not longitudinally ridged.

The *Palæophis porcatus* is characterised by the longitudinal ridges connecting the anterior with the posterior zygapophyses, by its broader and squarer neural arch ; but it has the two hæmal spines below like the other large species from Bracklesham.

PALÆOPHIS (?) LONGUS, *Owen*. Tab. XIV, figs. 35, 36, 37, 45, 46.

Vertebræ of a serpent agreeing in character with those of the London clay at Sheppy, but smaller, have been obtained by Mr. Colchester, from the sand of the Eocene formation underlying the Red Crag at Kyson or Kingston in Suffolk. In these, as in most of the trunk vertebræ of *Palæophis Typhæus*, the hypapophysis is a small sub-compressed tubercle at the under and back part of the body of the vertebra ; but there is no repetition of a smaller process at the fore part ; and no ridge is continued backward from the hypapophysis, as in the *Palæophis toliapicus*. The tubercle for the rib is single ; in *Naja* it is almost divided into two, the upper being convex, the lower moiety concave ; in the *Python* the upper half of the tubercle is convex, and the lower half concave, but the two facets are not marked off. In the fossil serpent from Kingston, as in the *Palæophis* from Sheppy and Bracklesham, the costal tubercle is simply convex. The chief characteristic of the Ophidian vertebræ from Kingston is the length and slenderness of their bodies, in which respect they exceed those of the *Palæophis toliapicus*, and resemble some of the existing tree-snakes (*Dendrophis*) with elongated vertebræ. The origin of the neural spine is limited to the posterior half of the arch (fig. 36) ; but the mutilation of the neural arch in the specimens I have yet had the opportunity of examining, prevents a prosecution of the comparison with any adequate advantage.

Genus—PALERYX.

The vertebræ of this extinct genus of Serpent (T. XIII, figs. 29-32, and figs. 37-38) differ from those of *Palæophis*, in the absence of the pointed aliform production of the hinder border of the neurapophyses, that border (fig. 29, *n*) descending from the neural spine to the posterior zygapophyses, with a convex curve as in most modern Serpents. The neural spine (*ns*) is low, the antero-posterior extent of its truncated summit exceeding the height. There is no point of bone extending outwards beyond the articular surface of the anterior zygapophysis (*z*), as in *Coluber, Vipera, Naja, Crotalus,* and *Hydrus;* in this character *Paleryx* resembles *Eryx, Python, Boa,* and *Palæophis.* The middle and posterior trunk-vertebræ of *Paleryx* differ from those of *Python* and *Boa*, and resemble those of *Eryx* in having a sharp and well-developed hypapophysial ridge (*h*) coextensive with the under surface of the centrum, and deepest at its posterior half; but the border here is gently convex, not angular as in *Eryx;* and the posterior border of the neurapophysis is less produced than in *Eryx;* the articular cup and ball are relatively larger, especially transversely; the cup is a full transverse ellipse, not circular as in *Eryx;* in this respect it resembles that of *Python* and *Palæophis.*

PALERYX RHOMBIFER. Tab. XIII, figs. 29-32.

In the vertebræ of this species the hypapophysial ridge (*h*) is sharp and well produced; the neural spine (*ns*) is rhomboid, not rounded off anteriorly; the zygosphene (*zs*, fig. 30) has the same relative vertical extent as in the Python. The diapophysial tubercle (*d*) is less elongated vertically than in *Python* and *Boa*, presenting proportions like those of the vertebra of the Eryx (fig. 22, *d*); the zygapophyses (*zz'*) are more pointed at their terminations. The figures 29-32, Pl. 2 represent the largest of the trunk-vertebræ upon which has been founded the genus and species above defined: they indicate a land Serpent of about four feet in length. They were obtained from the Eocene sand at Hordwell by Alex. Pytts Falconer, Esq., of Christchurch, Hants., to whose liberality I am indebted for the specimen figured.

PALERYX DEPRESSUS, *Owen.* Tab. XIII, figs. 37 and 38.

The smaller Ophidian vertebræ, indicative of the above species, agree in their generic characters with the foregoing; that is to say, in the shape and development of the hinder border of the neural arch in the relations of the diapophysis to the anterior zygapophysis, in the shape and size of the articular cup and ball of the centrum, and in the shape of the diapophysial tubercle for the rib. But the whole vertebra is more depressed; the hypapophysial ridge is relatively thicker and less produced; the zygosphene has much less vertical thickness, and there is a corresponding modification of the zygantra; the neural spine is relatively lower and of a different shape, having its anterior angle rounded off, and its posterior one more produced backwards.

As I have failed to discover modifications of the kind and degree above described in the dorsal or free rib-bearing vertebræ of the same species in any of the existing genera of Serpents, I am left to interpret such characters as indicative of a distinct species, probably of the extinct genus of Eocene Serpent above defined. The specimens of the vertebræ of the present species, which indicate a serpent of between two and three feet in length, were obtained from the Eocene sand at Hordwell Cliff, by Searles Wood, Esq., F.G.S., in whose museum they are preserved.

A few bones of serpents have been found in the superficial stalagmite, and in clefts of caves, in peat bogs, and the like localities, which bring their occurrence and deposition within the period of human history. None of these Ophidian remains, however, have offered any differences in size or other character from the corresponding parts of the skeleton of our common harmless snake (*Coluber natrix*). As yet no Ophidian fossils have been found in British fresh-water formations of the pre-adamitic or pleistocene period, from which formations the remains of the Mammoth, tichorrhine Rhinoceros, great Hippopotamus, and other extinct species of existing genera of *Mammalia* have been so abundantly obtained. Between the newest and the oldest deposits of the tertiary period in Geology, there is a great gap in England, the middle or miocene formations being very incompletely represented by some confused and dubious parts of the crag of fluviomarine origin in which teeth of a Mastodon have been found.

The deposits in which the remains of the large serpents of the genus *Palæophis* occur so abundantly, carry back the date of their existence to a period much more remote from that at which human history commences. Yet, as the strange and gigantic Reptiles that have been restored, and, as it were, called again to life, from times vastly more ancient, realise in some measure the fabulous dragons of mediæval romance; so the locality on our shore of the English channel in which the Eocene serpents have been found in most abundance and of largest size, recalls to mind, by a similar coincidence, the passage cited by an accomplished and popular historian, in his masterly sketch of the rise and progress of the English nation. "There was one province of our Island in which, as Procopius had been told, the ground was covered with serpents, and the air was such that no man could inhale it and live. To this desolate region the spirits of the departed were ferried over from the land of the Franks at midnight." (Macaulay's History of England, vol. i, p. 5.)

C AND J. ADLARD, PRINTERS, BARTHOLOMEW CLOSE.

TAB. XIII.

Ophidia.—Recent and Fossil Ophidian Vertebræ; nat. size.

Fig.

1—4. Vertebræ from the middle of the body of a *Python Sebæ*, twenty feet in length.

2,′ 3′. A rib of the same Python.

3″. Section of the proximal end of a rib of the same Python, showing the medullary cavity.

5—8. Vertebræ from the middle of the body of the *Palæophis typhæus*.

9—12. Corresponding vertebræ of a rattle-snake. (*Crotalus*.)

13—16. Corresponding vertebræ of a hooded-snake. (*Naja*.)

17—20. Corresponding vertebræ of a harmless snake. (*Coluber elaphus*.)

21—24. Corresponding vertebræ of an African *Eryx*.

25—28. Corresponding vertebræ of a sea-snake. (*Hydrus bicolor*.)

29—32. Corresponding vertebræ of *Paleryx rhombifer*.

33—36. Corresponding vertebræ of an Iguana.

37—38. Corresponding vertebræ of *Paleryx depressus*.

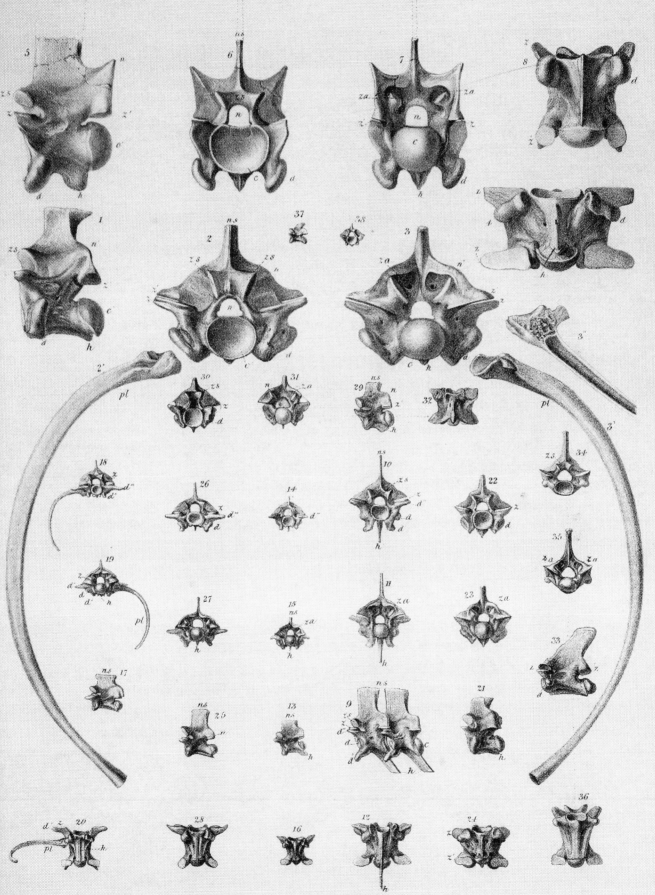

TAB. XIV.

OPHIDIA.—Fossil and Recent Ophidian Vertebræ; nat. size.

Fig.
1—3. A cervical or anterior trunk vertebra of a *Palæophis*.

4. A corresponding vertebra of a *Python tigris*.

5—6. A small vertebra of a Palæophis with one hypapophysis.

7—9. A larger vertebra of the same species.

10—12. A similar vertebra of the same species.

13—15. A small vertebra of *Palæophis porcatus*.

16—17. A large vertebra of Palæophis, longer in proportion to its breadth than

18—20. A type vertebra of *Palæophis porcatus*.

21. A vertebra of *Palæophis porcatus*.

22—24. A middle trunk vertebra of the same species of *Palæophis*, as figs. 1—3.

25. A small vertebra of a *Palæophis*, with an inferior ridge.

26. A large vertebra of   do.   do.

27. Two views of a vertebra of *Palæophis typhæus*, with the major part of the long neural spine preserved.

28. Two vertebræ of *Palæophis typhæus*, in natural articulation.

29—31. A vertebra of a *Palæophis*, of the compressed kind.

32—34. Two similar but smaller vertebræ of the same kind anchylosed, perhaps from the tail.

35—37. A trunk vertebra of the *Palæophis longus*.

38. Front view of the atlas vertebra of the *Python Sebæ*.

39. Side view of the same vertebra.

40. Side view of the axis vertebra of the same *Python*.

41. Front view of an anterior caudal vertebra of a *Python tigris*.

42. Front view of a middle caudal vertebra of the same Python.

43—44. Two views of a portion of the lower jaw of a lizard or sauroid fish.

45—46. The centrum of a vertebra of the *Palæophis longus*.

 With the exception of figs. 35—37 and 45—46, which are from Kingston in Suffolk, all the specimens of *Palæophis* figured in this plate are from Bracklesham, Sussex.

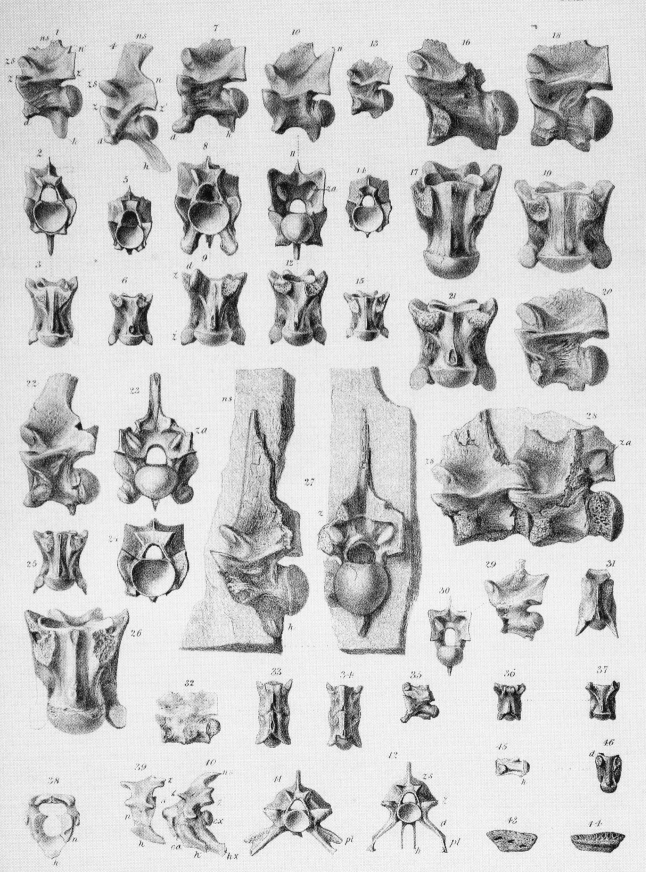

TAB. XV.

OPHIDIA.—*Palæophis toliapicus;* nat. size.

Fig.

1. A group of vertebræ, some of which show the long neural spine entire.

2. A group of vertebræ and ribs.

3. Five vertebræ in natural articulation with the diapophysis well preserved.

4. Under surface of four partially dislocated vertebræ.

5. Side view of the same vertebræ; natural size.

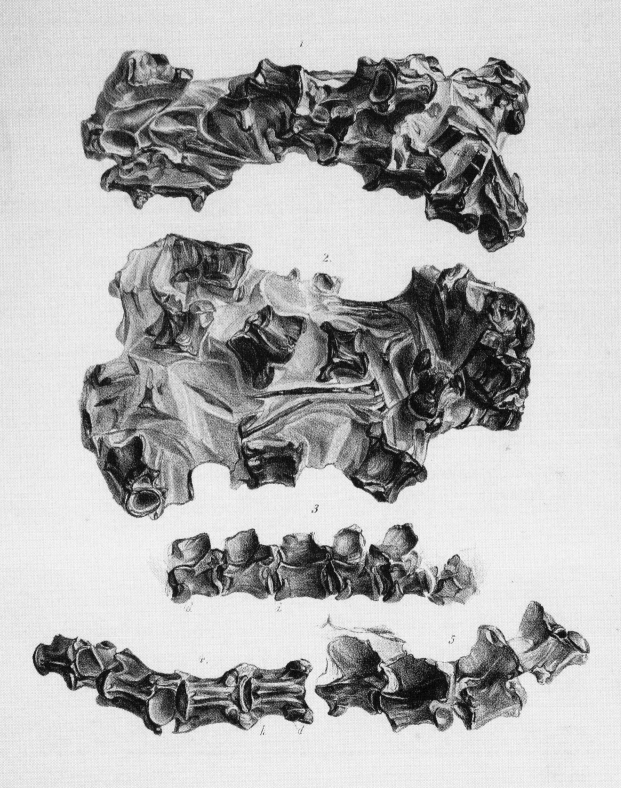

TAB. XVI.

OPHIDIA.

Fig.

1. Left side view of a chain of thirty trunk vertebræ of the *Palæophis toliapicus*, from Sheppy.

2. Right side view of the same vertebræ.

3. Under view of five vertebræ of the same chain, with the articular ends of some of of the ribs.

4. Portion of a group of Palæophis vertebræ from Bracklesham, showing the size and structure of the ribs; natural size

Fig.1

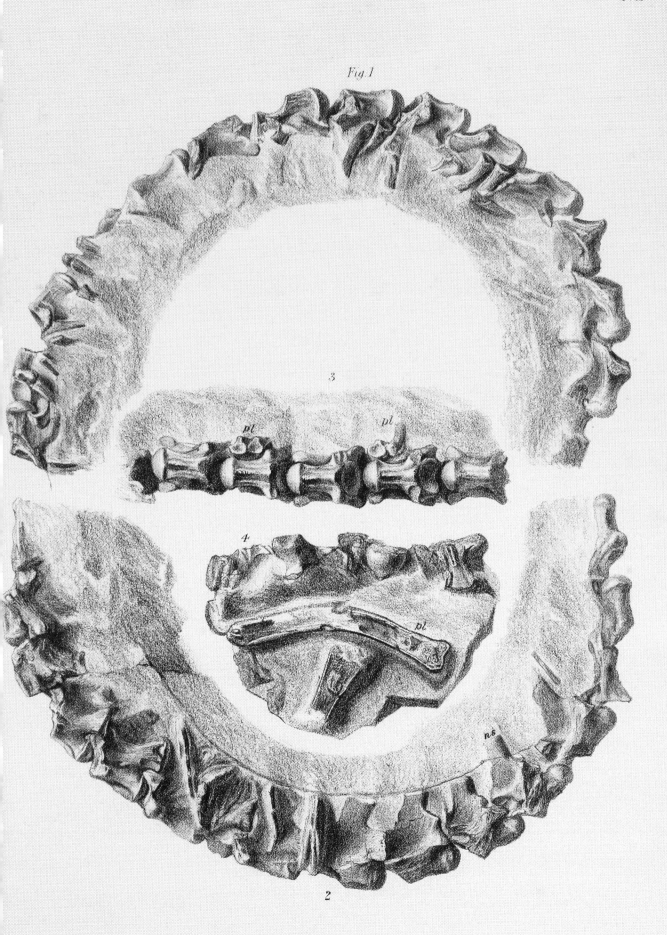

MONOGRAPH

ON

THE FOSSIL REPTILIA

OF THE

LONDON CLAY.

VOLUME II.   PART I.

PAGES 1—4; PLATES I AND II.

[CHELONE GIGAS.]

BY

PROFESSOR OWEN, C.B., F.R.S.,

FOREIGN ASSOCIATE OF THE INSTITUTE OF FRANCE,
ETC. ETC.

Issued in the Volume for the year 1880.

LONDON:

PRINTED FOR THE PALÆONTOGRAPHICAL SOCIETY.

1880.

PRINTED BY
J. E. ADLARD, BARTHOLOMEW CLOSE.

MONOGRAPH

ON THE

FOSSIL REPTILIA OF THE LONDON CLAY.

VOL. II.

RESTORATION OF *CHELONE GIGAS*, Ow., A SPECIES INDICATED BY A FRAGMENT OF THE FEMUR IN A FORMER MONOGRAPH.[1]

IN the ' Monograph on the Fossil Reptilia of the London Clay, Supplement to the Order Chelonia,'[1] the proximal end of the femur of a very large marine Chelonian is noticed and figured in Pl. XXIX, fig. 5, and the size of the species in proportion to the largest known existing Turtle, an individual of *Chelone mydas*, which weighed 150 lbs., is illustrated by a subjoined figure of the entire femur of that individual (fig. 5'). The breadth of the proximal end of this femur across the trochanters is barely two inches, the same dimension of the fossil is four inches ten lines.

In 1858 the British Museum obtained the upper portion of the cranium of a *Chelone* from the same formation and locality (Sheppey), corresponding in magnitude with the above-cited portion of fossil femur, and it was registered and labelled as of the *Chelone gigas*.[2]

In the present year, 1879, W. H. SHRUBSOLE, Esq., F.G.S., submitted to my examination an almost entire cranium and other portions of the skeleton of the same gigantic species of Turtle, which had been exhumed from the septarian modification of the London Clay at Sheppey. The following are dimensions of the fossil (Pl. I) and of the largest cranium of *Chelone mydas* in the Zoological Department of the British Museum.

---

[1] Palæontographical Society's Volume, issued for the year 1849, 4to., Pl. XXIX, fig. 5, 1850.

[2] ' Palæontology,' 8vo., 2nd ed., 1861, p. 317.

1

	Chelone gigas.		Chelone mydas.	
	inches.	lines.	inches.	lines.
Length . . .	18[1]	0	10	0
Breadth . . .	14[2]	6	7	0
Breadth across outside of tympanic articular end	13	6	6	3
Antero-posterior extent of outlet of orbit .	5	3	2	5
Vertical extent of outlet of orbit .	3	9	2	7
Height of nostril . . .	2	10	1	1
Breadth of nostril . . .	4	0	1	5

Notwithstanding an obvious flattening vertically with some distortion I incline to regard the skull of *Chelone gigas* as of relatively less vertical extent than that of *Chelone mydas*. The orbit is relatively larger and is longer in proportion to its vertical diameter. The slight downward extension of the upper border is near the fore part of the cavity in *Chelone gigas*, near the hind part in *Chelone mydas*. The nostril in *Chelone gigas* is broader in proportion to its depth, more especially towards its base.

The subtrenchant alveolar border of the maxillary is continued from beneath the orbit in an uninterrupted feebly convex line to the premaxillaries. There is no indication of the abrupt angular notch which produces the tooth-like process of the maxillary anterior to the orbit in *Sphargis*.

So far as the sutures can be traced on the upper expanse of the cranium (Plate I) they conform in the main with those of *Chelone mydas*.

The vaulted cavities roofed over by the parietals and mastoids are relatively lower and smaller in *Chelone gigas* than in *Chel. mydas*, and this irrespective of the degree of pressure which has somewhat affected the shape of the hinder outlets.

The superoccipital spine seems from the size of its broken base to have had the same relative degree of production beyond the foramen magnum in the gigantic fossil as in the recent Turtle.

The bilamellar base of the superoccipital element forming the keystone of the arch of the myelonal outlet has been slightly dislocated and pressed downward into the foramen magnum; but the exoccipitals retain their natural position as the side walls of that orifice. The part of the 'foramen magnum' (Pl. II, fig. 1) contributed by the basi-occipital is less prominent, more extensively concave, than in *Chelone mydas*, in which such concavity is represented by a central pit. The whole condyle is broader in propor-tion to its depth, and the basioccipital tract anterior to the condyle is relatively broader than in *Chel. mydas*.

So much as could be exposed by the mason's chisel of the palatal surface of the enormous cranium (Pl. II, fig. 3) accorded in the main with the configuration of the

---

[1] Part of the superoccipital spine has been broken away.
[2] Slightly increased by posthumous flattening.

same surface in existing Turtles. From the end of the occipital condyle to the posterior border of the *palatonaris* the extent is nine inches in *Chelone gigas*. The palatal opening is, as in *Sphargis*, more distinctly divided or indicative of the two nasal passages than in *Chelone mydas*.

The least breadth of the suturally united pterygoids dividing the lower temporal openings is two inches ten lines; in *Chel. mydas* it is one inch five lines.

The less perfect skull of *Chelone gigas* obtained in 1858 is represented chiefly by the upper wall, to the outer surface of which is cemented portions of the scapulo-coracoid arch. On clearing out the matrix from the inner or under surface of this specimen parts of the side walls of the cerebral cavity and of the orbits, with the alveolar border of the upper jaw, two inches and a half in advance of the orbit, were brought into view. The comparable parts confirm the affinity to the true Turtles, *Chelone*, as contrasted with the "leatherbacks" (*Sphargis*).

The extreme breadth of this cranium is . .	13 inches.
The preserved length . . .	15 ,,
The extreme breadth of the cerebral chamber	2 ,,

The portion of the alveolar border testifies, as in the better preserved specimen, to the Chelonial character.

The decisive test of the affinities of the Eocene gigantic representative of the marine family of the Order *Chelonia* is afforded by the portion of the petrified carapace and plastron associated with the later discovered skull. The Leatherback Turtles (*Sphargis*) have little of the carapace besides the normal unexpanded pleurapophyses (dorsal ribs) to undergo the conservative petrifying process.

The expanded horizontal plates from the summits of the neural spines and the corresponding expansions from the upper surface of the pleurapophyses, constituting the so-called neural and costal plates, have been preserved in eight portions of those modified segments of the carapace of *Chelone gigas* which have been recovered.

The neural plate answering to the 'sixth' of the entire carapace measures four inches in length and eight inches in breadth; the lateral halves slope with a feeble concavity from a slightly elevated medial rising as in the hinder plates of *Chelone subcristata*.[1] The costal sutural border makes a very low angle at about the same distance from the anterior border as in the corresponding plate of *Chelone longiceps*.[2] The marginal thickness of the plate is nine lines. The neural plates are relatively broader in proportion to their length than in any recent or hitherto observed fossil species of *Chelone*.

About one inch of the first neural plate and three inches of the seventh are preserved

---

[1] 'Monograph on the Fossil Reptilia of the London Clay,' Part I, *Chelonia*, issued by the Palæontographical Society, for the year 1848, p. 24, Pl. VIII, fig. 1, 1849.

[2] Ib., ib., p. 16, Pl. IV, fig. 2, *s* 6.

in connection with the intervening neural plates, which are entire save some abrasion of the outer surface of the hinder ones and slight mutilation of the sutural borders.

The costal plates, of which the proximal parts of four are preserved on the right side, have been slightly dislocated from their marginal sutures with the neural plates by super-incumbent pressure. A longitudinal extent of eight and a half inches is preserved of the left sixth costal plate.

On the under part of the mass of petrified clay to which the above-described portion of carapace is cemented the expanded sternal portions of the pair of cora-coid bones are preserved converging to the median line; a portion of the xiphoid production of the entosternum extends along the medial line in advance of the broad ends of the converging coracoids. This is the chief recognisable part of the plastron here preserved, but it adds to the testimony against the *Sphargis* affinity, in which genus the entosternal does not send backward such xiphoid process.

The breadth of the sternal end of the coracoid of *Chelone gigas* is six inches; that of the scapular end shown in another portion of the matrix is two inches and a half; the total length of the coracoid is one foot five inches.

PLATE I.

Chelone gigas, natural size.

Upper view of the skull.

From the London Clay, Isle of Sheppey. In the Collection of W. H. Shrubsole, Esq., F.G.S., of Sheerness-on-Sea.

PLATE I

The material originally positioned here is too large for reproduction in this reissue. A PDF can be downloaded from the web address given on page iv of this book, by clicking on 'Resources Available'.

Chelone gigas, one third natural size.

Fig.
1. Side view of the skull.

2. Under surface of the skull.

3. Back view of the cranium.

From the London Clay, Isle of Sheppey.   In the Collection of W. H. Shrubsole, Esq., F.G.S., of Sheerness-on-Sea.

PLATE II.

3

1

2

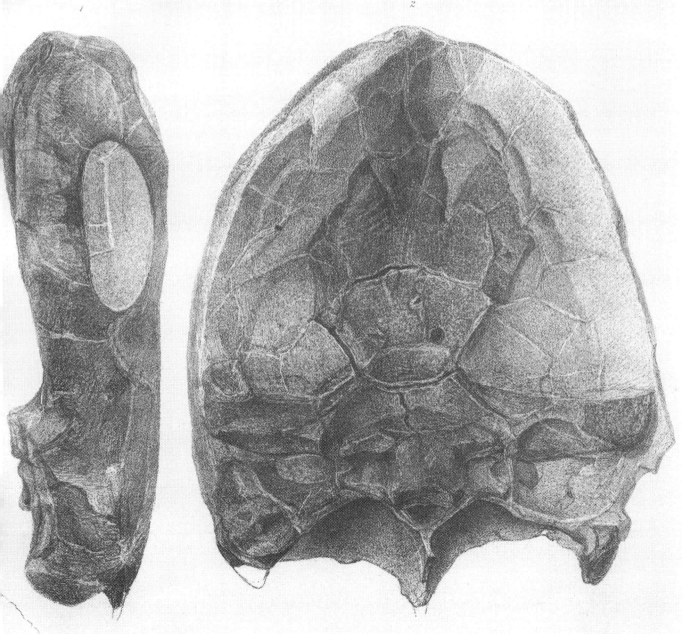

rom nat on stone by J.Erxleben.

Hanhart imp.

Printed in the United States
By Bookmasters